# The Wright Company

# THE
# Wright
# Company
*from invention* **I** *to industry*

EDWARD J. ROACH

Ohio University Press     Athens

Ohio University Press, Athens, Ohio 45701
ohioswallow.com
© 2014 by Ohio University Press
All rights reserved

Printed in the United States of America
Ohio University Press books are printed on acid-free paper ∞ ™

                                                                                    .

23 22 21 20 19 18 17 16 15 14 14     5 4 3 2 1

Library of Congress Cataloging-in-Publication Data available upon request

For Naomi

# CONTENTS

# ILLUSTRATIONS

# PREFACE

More than a century after his untimely death from typhoid fever, Wilbur Wright remains a famous man. His younger brother Orville, who died an elderly man in 1948, is also internationally famous. During the summer of 2012, Wikipedias in eighty-five different languages, from English and German to Kalmyk, Papiamento, and Võro, contained articles (of varying lengths) about the brothers: their invention of the first powered, heavier-than-air, human-controlled airplane and their first flights on the windy North Carolina coast in December 1903.[1] Their lives and achievements after they returned from Kill Devil Hills to their hometown of Dayton, Ohio, are less well known. Indeed, their invention of the airplane in Dayton is less well known than their first flights with it on the Outer Banks. Dayton, though, was where they designed and fabricated the gliders and airplanes they took to North Carolina, and through hundreds of test flights at Huffman Prairie Flying Field, outside the city, they turned their curiosity of 1903 into the practical aircraft of 1905.

Dayton was also the city where they decided to base production operations of the Wright Company, the brothers' attempt to commercialize their invention and sell it to government clients and aviators across North America. The brothers' business venture, which they established with the assistance of a group of prominent New York capitalists, remained in Wright hands from 1909 to its sale to a separate group of New York–based industrialists, in the autumn of 1915, by Orville Wright. The brothers have been well studied, with prominent biographies by Smithsonian curator Tom Crouch and Fred Howard of the Library of Congress leading the way. Studies on the Wright Company's aviation schools at Montgomery, Alabama, and at Huffman Prairie have also appeared in recent years. There are several works on Wright competitor Glenn Curtiss, including John Olszowka's excellent 2000 doctoral dissertation, and even on the history of the Burgess Company, which built licensed Wright-model aircraft for a few years. The early aviation industry as a whole has received some scholarly attention, of varying quality, in the past few decades. Curiously, Dayton—the City of a Thousand Factories and, during the Wrights' era, the home of such major companies as Barney and Smith and National Cash Register—has received almost no scholarly attention save

Judith Sealander's 1988 study of business progressivism and some essays and book chapters on the influence of the local branch of Eugene V. Debs's Socialist Party of America. Even studies of period business and labor history, while providing regional and national context for the Wright Company era, generally ignore Dayton. Little of this work about Dayton and the Miami Valley has focused on the company established by the inventors of the airplane, its short and difficult existence in Dayton, and its effects on aviation.[2]

I attempt here to begin a conversation—rooted in the extant papers of the Wrights, the Wright Company and its leaders, period trade journals and newspapers, and the existing scholarly literature—to start to fill in this gap. The Wright Company is not a particularly easy company to research. Its extant papers are incomplete (their quality declines markedly after 1913) and spread among different institutions, with the Museum of Flight in Seattle and the Library of Congress having the largest collections. Its financial records are spotty, and no good records of just who worked for the company exist. Dayton's industrial history, of which the Wright Company was only a small part, and the work of the Wrights between 1903 and 1909 to make their airplane a commercial product and to obtain and enforce patents on their work, deserve careful study.[3]

This is but a piece of a larger literature concerning the Wrights in the years after their North Carolina flights. Those interested in the specifications of Wright airplanes should turn to Richard Hallion's reference guide and to the appendix of the collection of Wright papers edited by Marvin McFarland. Paul Glenshaw has briefly addressed both the exhibition department and Roy Knabenshue's airship career in the Smithsonian's popular magazine *Air and Space*. John C. Edwards covers aspects of both the exhibition department and the Huffman Prairie flight school in his work, while Julie Williams provides a study of the 1910 Wright Company school at Montgomery, Alabama. Nevertheless, the Dayton-Wright Airplane Company, formed in 1917, has yet to be thoroughly studied. All these works have contributed to what you hold in your hands.[4]

# ACKNOWLEDGMENTS

This book began as a revision of an unsubmitted 1989 National Historic Landmark nomination for the Wright Company's factory buildings in west Dayton; over the past few years, it has evolved into a much larger project with a variety of supporters and assistants. The National Park Foundation provided a grant that supported the acquisition of a variety of primary and secondary sources, especially from the Wright Company collection at the Museum of Flight and from Frank Russell's papers at the University of Wyoming's American Heritage Center. Intern Emily Tragert diligently copied dozens of files in the Wright brothers and Grover C. Loening collections at the Library of Congress, while intern Andrew Hall conducted essential, if tedious, research in the microfilmed editions of the unindexed Dayton newspapers at the Dayton Metro Library to find coverage of the Wright Company. Nancy Horlacher, the local history librarian at the Dayton Metro Library, graciously allowed me to borrow some of the library's extra copies of the Dayton city directories from 1909 through 1915, enabling me to page through them for Wright Company employees from the comfort of my office. Dawne Dewey and her staff at Wright State University's Special Collections and Archives were as helpful as ever. Susan Roach, my mother, provided essential interlibrary loan resources in the early stages of the project and support throughout, as did my father, James Roach. Dayton Aviation Heritage National Historical Park, which had its boundary expanded in 2009 to include the company's former factory buildings, provided a very supportive (if sometimes noisy) environment for the production of this book. I especially thank the park's superintendent, Dean Alexander, and the former chiefs of education and resources management Ann Honious (1995–2009) and Noemi Ghazala (2010–12) for their support. I hope this book will serve the staff and constituency of the park and its partners in the coming years. The history departments at Moravian College and Indiana University of Pennsylvania helped build the foundation that produced this book. Juliet Burns, our level-9 Siamese Residential Agitator, made certain that proper routines were followed. Most important, Naomi Burns, a skilled copy editor, has put up with the gestation of this work and my dinner table discussions of Frank Russell and citation style. Her services and support throughout are incalculable.

FIGURE 1.1. The vacant former Wright Company buildings in 2012: 1 (*left*, built in 1910) and 2 (*right*, built in 1911). *Photo by author*

# INTRODUCTION

In west Dayton, Ohio, an empty factory complex quietly stands. Wedged between U.S. Route 35 and West Third Street, two of Dayton's major roads, the site is similar to many other former industrial sites throughout the Rust Belt, awaiting redevelopment and new investment. The site, though, contains two buildings built when Dayton was an industrial powerhouse, a city famous for its factories. These buildings, the former factory of the Wright Company, were the first buildings in the United States built specifically to house an incorporated airplane builder. Vacated by the Wright Company in 1916 and used as part of an automobile parts plant into the 2000s, the buildings today are monuments to Wilbur and Orville Wright's attempt to turn their invention into a profitable commodity, an attempt they found difficult to realize.

Aviation as a business in the United States changed greatly in the years between Wilbur Wright's demonstration flights at Le Mans, France, in 1908, and the start of the First World War, in 1914. In 1908 it barely qualified as an industry. The men—and initially they were all men—who pursued flight in the United States and Canada generally did so with aircraft they personally built and modified. Wilbur and Orville Wright and Glenn Curtiss experimented on single machines built in small, informal settings either with personal funds or through the assistance of a wealthy benefactor (as with Alexander Graham Bell's role with Curtiss and the Aerial Experiment Association), not with capital raised through a sale of company stock. By 1908 the Wrights and Curtiss, now satisfied that their airplanes were practical vehicles, looked to profit financially from marketing their wares. To begin to bring airplanes before the public as the period's automobile makers were commercializing their products, Wilbur and Orville Wright and Glenn Curtiss all required outside investment since on their own they did not have sufficient capital for commercial-scale airplane production. In 1909, after gaining capital from outside investors, both parties opened small airplane factories in their hometowns and started selling their models to governments and private buyers. Airplane manufacturing was a small, skilled craft in its first years. Its status as a small industry changed with the coming of the First World War, both in North America and Europe. By 1914 thousands of workers at companies in

Great Britain, Germany, and France—joined over the next four years by hundreds of people working in North America—built standard-model airplanes on variants of the assembly line for the air forces of the Allies and the Central Powers. European airplane production greatly outpaced that of the United States before the coming of the war. The U.S. industry, ensconced in a country distant from the arms races in Europe and greatly affected by patent infringement lawsuits, developed fitfully in the years before the war.

The Wrights' efforts at capitalizing on their invention were at the center of this fitful development. Vigorously defending their primacy and intellectual property was the Wright Company, formed by the brothers and a group of investors in 1909 to market their invention to North Americans and to prosecute infringements of the Wrights' 1906 patent (which the brothers assigned to the company upon its formation). Incorporating the brothers' fame in its name, the Wright Company produced thirteen different models of airplanes and served as an introduction to industrial aviation for individuals who later became prominent in aviation manufacturing such as Frank H. Russell (1878–1947) and Grover C. Loening (1888–1976). But the Wrights were engineers and inventors, men who previously owned and operated a small printing shop and a bicycle sales and repair business. Neither brother had any experience in running a company co-owned with a group of stockholders who wanted the firm in which they invested to grow to be the dominant airplane maker in the United States. Yet they resisted implementing suggestions for corporate growth from executives and managers with actual experience in running larger businesses. Moreover, the primacy the Wright Company gave to patent litigation over business development and technological innovation and Orville Wright's lack of interest in corporate management after the 1912 death of his brother, confidant, and business partner, with whom his work was closely intertwined, caused it to remain a small (if well-known) operation, even after Orville Wright sold it to a group of New York–based industrialists in the autumn of 1915.

As a specialized batch producer, the Wright Company is representative of the starting point of the transition of the aviation industry from craft production to assembly lines turning out thousands of airplanes each year. Its story shows that a famous name is insufficient to ensure a company's success and that a company marketing a new product—especially in times of recession—needs to combine innovative products with competent front-office management. Still, the prominence of its two presidents as the inventors of the airplane and the attempts it made at controlling the industry through patent infringement litigation gave the Wright Company a place in early aviation greater than the small number of airplanes—approximately 120 between 1910 and 1915—that its workers built in Dayton would otherwise indicate. When Orville sold the Wright Company, in 1915, he had realized his and his brother's dream of turning aviation from an oddity into a practical enterprise.

# I

## "We Will Devote . . . Our Time to Experimental Work"

*Creating the Wright Company*

In 1905, nearly two years after their first four flights on the North Carolina coast, Wilbur and Orville Wright succeeded in developing what they deemed a practical airplane—one in which a pilot could take off and land repeatedly as long as it maintained a sufficient fuel supply. On the fifth of October, Wilbur Wright flew nearly twenty-four miles (thirty-nine kilometers) in thirty-nine minutes, circling Huffman Prairie, outside Dayton, Ohio, and landing only when his airplane exhausted its fuel. Now satisfied with their invention's practicality, and hoping to begin to turn a profit on it, the Wrights decided to end their test flights and turn their attention to procuring patent protection for their airplane—which they received on 22 May 1906—and to marketing it. Neither Wilbur nor Orville would be a pilot for more than the next two years while they attempted to interest the U.S., British, German, and French governments in their airplane. They first approached governments as likely customers, as they anticipated that military uses for aviation, particularly in scouting enemy positions, would drive sales, at least initially. The expense of the new technology—the Wright Company offered its Model B for $5,000 in an era when the U.S. per capita income was just over $338—made it unaffordable for most private individuals.[1]

Politicians and bureaucrats in Europe and North America were slow to see value in the Wrights' invention, and it took the brothers several years to gain purchase contracts from the U.S. Army Signal Corps and from other governments. Fulfilling the terms of those contracts, though, created significant public interest in aviation, as governments usually required demonstration flights before finalizing contracts. Even with this interest, though, the Wrights were not eager initially to join the ranks of the tycoons of industry. In a November 1907 conversation in Paris with Hart O. Berg, who became the brothers' European business agent, Wilbur noted that he did not "want to build up a big business" but instead hoped "to get the greatest amount of money with as little work" as possible. In 1907 he stated the brothers' motto: "not 'lots of machines,' but 'most money.'"[2]

During the next two years, the brothers' interest in commercializing their work changed. They publicly exhibited the capabilities—and dangers—of their airplane to audiences in North America and Europe. Orville Wright twice flew at Fort Myer, Virginia, as part of a contract to deliver an airplane to the U.S. military. A crash during his first visit to Virginia, in September 1908, seriously injured Orville and killed his passenger, army lieutenant Thomas Selfridge, the first person to die from injuries sustained in an airplane accident; Orville returned to Virginia and fulfilled the terms of the contract the next summer. Meanwhile, Wilbur Wright awed crowds in France in fulfillment of a contract between the Wrights and La Compagnie Générale de Navigation Aérienne, formed in December 1908, to manufacture and sell Wright airplanes in the French market. The Wrights also established a manufacturing company in Germany (Flugmaschine Wright-GmbH) and contracted with Short Brothers, a London firm, to manufacture their airplanes in Great Britain. Having arranged to sell their airplane in several of western Europe's larger economies, the Wrights began to consider establishing a corporation to market their airplanes in the United States.[3]

In 1909 the Wright brothers were at the height of their influence. The balding Wilbur, then forty-two, and the slightly less bald (but mustachioed) Orville, thirty-eight, had decades of engineering and mechanical work behind them. Growing up in a household both mechanical and intellectual, Milton and Susan Koerner Wright's two youngest surviving sons (twins, a boy and a girl, died shortly after birth in 1870) had built printing presses and bicycles before their attention shifted to flight after the death of German aviator Otto Lilienthal in 1896. Their mechanical abilities came from their mother, whose father was a carriage maker and farmer in Loudoun County, Virginia; though a skilled writer and editor, Milton Wright, who kept the plain, simple Anabaptist attire and beard without mustache his entire adult life, "was one of those men who had difficulty driving a nail straight." But even if Milton's ability with tools was suspect, he remained an influential figure in his sons' lives until his death, in 1917 (Wilbur died in 1912). The brothers' first career, as job printers, arose from the reason that the Wright family came to Dayton: the United Brethren Printing Establishment, which published the denomination's newspaper, *The Religious Telescope*, a weekly to which Milton Wright was elected editor in 1869. To print handbills, business directories, and letterhead (and a short-lived newspaper, *The Tattler*, for Orville's former high school classmate, Paul Laurence Dunbar), the brothers designed their own printing press, though they also used a commercially produced Prouty press for most of their smaller jobs. As the brothers became engulfed in the bicycling craze that followed the development of the safety bicycle in the 1880s, they decided to open their own bicycle sales and repair business and in 1895 began to build their own brand of bicycle in their small shop, half a block north of their home. Between 1895 and 1904, nearly three hundred Van Cleve, St. Clair, and Wright Special models came out of the brothers' workshops, first at 22 South Williams Street and, after 1897, at 1127 West Third Street. But both in the printing business, which the brothers left in 1899, selling their equipment to Thomas and Marion Stevens of Dayton, and in bicycling, in which the Wright Cycle Company technically remained a venture until 1908, the Wright brothers were insignificant operators. While they occasionally employed a few friends to help with printing, and though their famed mechanic, Charles E. Taylor, first gained employment with the brothers through the bicycle shop, their businesses were dwarfed by the printing operations of Dayton's major newspapers—the *Daily News*, the *Journal*, and the *Herald*—and by George P. Huffman's Davis Sewing Machine Company, which later became Huffy Manufacturing, still a producer of popular bicycles. But the Huffman family retains a connection to the Wrights, as George's banker brother, Torrence, allowed the Wrights to use a cow pasture he owned just outside Dayton as their local testing field. Huffman Prairie Flying Field, which in 1917 became part of Wilbur Wright Field (itself now a part of Wright-Patterson Air Force Base), is still famous among aviators. The Wrights, as printers, as bicycle makers, and as pioneer aviators, had always worked closely together, with limited hired help. They would find it difficult to transition from their small business background to the world of incorporated entities.[4]

Though they had been first in powered, controlled flight, the Wrights were not the first people in the United States to establish an airplane company. That distinction belonged to Glenn H. Curtiss and the shady Augustus Moore Herring, who formed their ill-fated Herring-Curtiss Company in Hammondsport, New York, in March 1909, with an official market capitalization of $360,000. While the Wrights sued Herring-Curtiss that summer for patent infringement (launching a court case that eventually ended in an appellate court decision in favor of the Wrights in 1914), Herring-Curtiss declared bankruptcy in 1910, after Curtiss and the rest of the company's board realized that Herring had not actually received any of the patents he claimed to own, patents that provided the company with much of its supposed value. Glenn Curtiss then organized the Curtiss Aeroplane Company to produce and market his airplanes. Even though the Wrights were second in incorporating, the press reported that they seemed poised for success. As a result of their demonstration flights, claimed the *New York Times* in May 1909, the Wrights had "more than eighty orders for airships to be used in the United States" and had received inquiries from parties in Iceland, Iran, and China. However, the small size of their "factory"—at the time, their Dayton bicycle shop—made it impossible for them to fulfill the purported orders. Interestingly, the ability of financing such production did not seem to be of much concern for the Wrights; the *Times* claimed that they turned down an investment offer from "well-known New York capitalists" to incorporate and enlarge their facilities. But the brothers clearly maintained interest in commercializing their work domestically even as they flew in Europe in the spring and in Virginia in the summer—they were just waiting for the right opportunity to arise.[5]

The Wrights spent most of their lives in a city famous for its factories. While the Wright Company was the sole incorporated aviation-related company in Dayton in 1910, its airplanes were just one of the many products of the workers in the city's plants. Called the City of a Thousand Factories by the *Boston Evening Transcript* in 1913, industrial production dominated the economy of Dayton during most of the twentieth century. In 1910 the city—by population the forty-third largest in the United States, fourth largest in Ohio, and the county seat of Montgomery County—was home to 116,577 people, an increase of more than 31,000 residents since 1900. Its residents were mostly native-born whites (83.9 percent); immigrants (11.9 percent) and African Americans (4.2 percent) accounted for much smaller portions of the population. Germans were the largest immigrant group in the city (4.2 percent of the overall population), with ethnic Hungarians constituting nearly 3 percent and people from the Russian Empire forming 1 percent. But Dayton was not a particular magnet for immigrants who decided to remain in Ohio. New arrivals to the United States made up significantly larger proportions of the populations of Cincinnati (15.6 percent) and Cleveland (35 percent), Ohio's two largest cities. The proportion of what the Census Bureau's *Statistical Abstract* labeled in the polite terms of the day the city's Colored population was slightly higher than the overall African American population in the state, where blacks made up 2.3 percent of the state's 4.76 million residents, but Dayton's population of native-born and foreign-born whites reflected the proportion of those groups in the state's overall population, which the 1910 census determined to be over 85 percent. Single-family homes, not apartment blocks, dominated Dayton's neighborhoods.[6]

Industrial development had a long history in Dayton. During the nineteenth century, Dayton developed as a major industrial center largely through the growth of the Barney and Smith Car Company (which built passenger and freight cars for railroads throughout the world and employed several thousand people at its apex) but also through the development of firms that supplied goods and materials to Barney and Smith. By the turn of the century, Barney and Smith, while still a major employer in the city (with over thirty-five hundred employees as late as 1909), was in decline as the result of a shift from wooden to steel railroad cars and of poor leadership.[7]

Fortunately, when Wilbur and Orville Wright sat down with their investors in New York to create the Wright Company, their hometown was not a one-industry city economically dependent on the fate of Barney and Smith for its survival. Other companies supplemented and surpassed the railroad carmaker's impact on the city and the nation. The National Cash Register Company (NCR), established in 1884, emerged as Dayton's largest manufacturer during the 1890s through aggressive marketing and employing new methods for mass producing cash registers (the invention of Dayton saloon-keeper James Ritty and his brother John in 1879)—methods implemented by

9

ORVILLE WRIGHT FLYING OVER CITY
IN HIS AEROPLANE —— DAYTON, O.

FIGURE 1.1. A postcard showing Orville Wright flying a Model B over downtown Dayton, 22 September 1910. *Courtesy of the Dayton Metro Library, Wright Brothers Collection*

its founder, John H. Patterson (1844–1922)—and through the quality of the products produced by the company's workers. Under Patterson's leadership, NCR became an early model of the modern corporation. Two of Patterson's protégés at NCR, Charles Kettering and Edward Deeds, established Dayton as an important maker of automobile parts with the formation of the Dayton Engineering Laboratories Company (Delco) in 1909, while hundreds of workers built John Stoddard's Stoddard-Dayton and Pierce Schenck's Speedwell automobiles at factories in the city. Indeed, Speedwell would rent space to the nascent Wright Company while it built its own factory. Industrial employment was plentiful in Dayton—if one was a white male. The city's major companies employed white workers exclusively in all but the most menial of jobs, as poet Paul Laurence Dunbar discovered when he attempted to find work as a writer in the 1890s but could get a job only as an elevator operator. No African Americans worked at NCR, where Patterson had fired the only African Americans he employed—the company's janitorial staff—in 1897. Patterson was famous (or infamous) for his business welfare programs, which included worker cafeterias (Patterson held esoteric nutritional beliefs; all beverages a worker drank were to be hot, and his cafeterias never served ice water), free employee baths (important when not every worker lived in a home with running water), beautifying his employees' neighborhoods, and providing them a country club (today's Community Golf Club, not the course that hosted the 1969 PGA championship) for recreation—and his white handlebar mustache. He asserted that every NCR employee should have the opportunity to advance in the company on merit but that, since the segregated social system would not permit blacks to supervise whites, no African Americans should work for his firm. Patterson was not W. E. B. DuBois; he was not going to challenge the system. And neither was the Wright Company. There is no evidence that the Wright Company ever employed African Americans or trained any as pilots at its Huffman Prairie flight school. Racism pervaded early aviation, and Bessie Coleman—the first African American to earn a pilot's license—had to travel to France to do so in 1922.[8]

Educational institutions in Dayton reflected the needs of an industrial city. While literacy was widespread—census takers found only 2.3 percent of the overall population over the age of ten to be illiterate, though 10.9 percent of African Americans could not read or write—few people graduated from high school or possessed college degrees. In an era when only 13.5 percent of the U.S. population over twenty-five had graduated from high school and only 2.7 percent from college, Dayton had but two public high schools (Steele and Stivers, to which twenty-three elementary schools supplied students) and only small, religiously oriented, postsecondary educational institutions. Segregated high schools came to Dayton in 1893, shortly after Paul Laurence Dunbar's graduation from Central High School, in 1890. Dunbar could be in Orville Wright's class at Dayton's old Central High because too few African Americans were then enrolled for secondary studies to make a separate black high school financially feasible. Dayton was not famed for liberal arts education. St. Mary's Institute (renamed the University of Dayton in 1920), operated by the Roman Catholic priests and brothers of the Society of Mary, enrolled approximately 390 students in its collegiate and high school classes, while Bonebrake Theological Seminary, an institution that trained United Brethren ministers and missionaries (founded in 1869 as Union Biblical Seminary), enrolled sixty-five. Milton Wright, who gave the seminary its original name, was part of its first executive committee. Three Dayton business colleges educated stenographers, accountants, and clerks for office jobs in local industry.[9]

Such secretarial fields, as well as teaching, nursing, and domestic service, along with occupations connected with sewing and cigar making, made up the employment opportunities generally available to women in Dayton. After her graduation from Oberlin College, in 1898, Katharine Wright, Orville and Wilbur's sister, taught Latin at Steele High School until she left the profession, rushing to Orville's side after his accident at Fort Myer, in 1908. Among the Dayton working class, women who worked outside the home usually worked in the city's clothing and cigar-making establishments. In 1910, Dayton had no significant textile mills, though nearly a thousand women worked as dressmakers or seamstresses. More than thirteen hundred women worked in the cigar industry, which also employed 261 men in more than forty separate factories. Women accounted for 84 percent of the city's teachers, 96 percent of its nurses, and 83 percent of its servants, the largest single category of employed women. While most women found their work circumscribed to these particular trades, African American women found their job opportunities even more limited. Discrimination was rife in Dayton, a city just fifty miles north of the Ohio River (and the Kentucky border), a city whose district notorious copperhead Clement Vallandigham represented in Congress from 1858 to 1863. Of the 893 black women employed in Dayton in 1910, 703 worked either as servants (502) or as launderers (201), as did Lottie Jones, who washed clothes for the Wright family for decades and eventually acquired their home at 7 Hawthorn Street. Henry Ford purchased the house from her, not from Orville Wright, before he moved it to his Greenfield Village museum in Michigan. But Jones worked for the family; the brothers paid her from their personal joint checking account. Neither she nor any other African American women are known to have worked for the Wright Company.[10]

Daytonians lived in a fervid political environment during the early 1910s, though Wilbur and Orville Wright generally avoided associating their names with politicians of any stripe. James Cox, the former Ohio governor and 1920 Democratic candidate for president and an acquaintance of the Wrights, wrote to a journalist friend in 1939 that Prohibition was the only topic on which Orville, a dues-paying member of the Anti-Saloon League, was a radical. The brothers grew up in a solidly Republican family. Over the decades, Milton Wright almost always voted the Republican ticket (though not blindly; in 1905 he voted for the successful Democratic candidate for Ohio governor, John M. Pattison, and refused to vote for some Republican candidates in 1906 as a protest against "Hearstism and Campbellism"—the views of those Republicans were too similar in his mind to Democrats William Randolph Hearst of New York and James Campbell of Ohio). But it was not battles between Republicans and Democrats that featured in the Dayton press. In 1913, after a concerted campaign led by National Cash Register president John H. Patterson, and after the incumbent mayor and city council remained all but invisible during and two months after the devastating flood

FIGURE 1.2. Wilbur (*left*) and Orville Wright on the porch of their 7 Hawthorn Street home in Dayton, 1909. *Courtesy of the Library of Congress*

that spring, city voters approved a plan to change the city's government. They decided to replace the strong mayor and a council with members elected by ward with a weak, officially nonpartisan council elected citywide on which the mayor would be only one of the five votes, a first among equals. While the mayor and council would serve a legislative function, day-to-day city operations were now the responsibility of an appointed city manager to whom city departments and employees reported. Both the Republican and Democratic parties acquiesced to the new system (Edward E. Burkhart, a Democrat and Dayton's former mayor, was part of the committee led by Patterson that put together the proposal for the new government), but Dayton's vocal and active Socialist Party vigorously opposed it, asserting that its candidates would not be competitive without wards in which to run or identification with a specific party. The Socialists were unable to revoke the system, however, and the weak mayor–city manager system remains in place today. Though neither the

Wright family nor the Wright Company had any significant involvement in the debates over the city's style of government, Orville Wright was on friendly terms with many of the campaign's leaders, especially NCR officer Edward A. Deeds, who later secured the Wrights' 1905 airplane for display in Dayton at his Carillon Park.[11]

Dayton was a bustling, thriving industrial center in 1909, and Wilbur and Orville Wright, confirmed bachelors living with their father and sister on the west side of town in the home in which they grew up, had ensured that they were solidly part of its middle class, particularly of its petite bourgeoisie. While they had owned their own businesses, few employees reported to them, and neither Wright and Wright Job Printers nor the Wright Cycle Company issued any sort of stock, was incorporated, or was overseen by a board of directors, facts that would haunt the brothers as they struggled to run the incorporated entity that became the Wright Company. Like many of their colleagues and competitors, including Glenn Curtiss, who left school after eighth grade, and Glenn Martin and A. Roy Knabenshue, high school dropouts, neither Wright brother received a high school diploma, and neither attended college. While many of the important early aviators did not achieve high school or college diplomas, most of their investors and business managers received much more formal education, often at elite institutions, and the Wrights' relationships with college graduates—especially the company's two managers, Frank H. Russell and Grover C. Loening, alumni of Yale and Columbia, respectively, were especially strained. Most of the Wright Company's directors were also college men: Robert Collier was a Georgetown graduate; vice president Andrew Freedman earned his degree from the College of the City of New York (now New York University); De Lancey Nicoll, an alumnus of Princeton and Columbia Law; and August Belmont, Harvard. Katharine Wright would be the only Wright sibling to graduate from college (Oberlin, 1898), though Bishop Milton Wright and Susan Koerner Wright both attended Hartsville College, in Indiana, and Milton taught for a time at the United Brethren institution.

College graduates or not, Wilbur and Orville Wright grew up with parents who supported education and critical thinking, their father a leading editor and bishop of the Church of the United Brethren in Christ and one of the founders of Union Theological Seminary, and a mother who possessed technical abilities foreign to her husband. But the brothers' technical and intellectual gifts did not include deep understandings of the intricacies of the financial world or experience in running an incorporated business with a board of directors with fiduciary interests in the enterprise. By the end of 1909 the brothers' French and German companies were running, and some income from sales of their airplanes on the continent was filtering into their pockets—as was money from the U.S. government for the fulfillment of their initial contract with the Signal Corps. Wilbur and Orville Wright could live off this money for a time, but it was not enough to capitalize a viable U.S. company. To take the next step toward making the Wright name dominant among private aviators and armies, the Wright brothers needed help. Dayton was an industrial city, not a hub of finance, and such major local employers as NCR, Barney and Smith, and Dayton Malleable Iron were long-established concerns. There were no Belmonts, Vanderbilts, or Freedmans in southwestern Ohio, and the brothers could not rely on local bankers for significant assistance beyond Torrence Huffman providing their proving ground. To demonstrate the national importance of aviation and the leading role of their own airplanes in the young field, they looked beyond the Miami Valley to the titans of Wall Street for investment capital. Still, Dayton was home, not New York, and the Wrights had no interest in moving there from their hometown. Instead, they would try to use Gotham's money to make Dayton the capital of the embryonic U.S. airplane industry. Wilbur and Orville Wright were embarking on a completely new journey.

# Bringing an Aeroplane Factory to Dayton

Nineteen hundred and nine, like most years, was full of important events. For the Cincinnati-born lawyer, judge, and former secretary of war William Howard Taft, it was his first year in the White House. The U.S. Mint introduced a cent bearing the portrait of Abraham Lincoln, the first U.S. coin depicting a real person. Ida Wells Barnett, W. E. B. DuBois, and others formed the National Negro Committee, the immediate predecessor of the National Association for the Advancement of Colored People. And it was a banner year for the Wrights. Early in the year, the two brothers conducted dozens of successful demonstration flights on the edge of the Pyrenees, in Pau, France, thrilling European notables, including King

Alfonso XIII of Spain. The Smithsonian Institution awarded the brothers its new Langley medal for aeronautics (though the award ceremony for the brothers would wait until 1910). In April they flew in Rome before King Victor Emmanuel III of Italy. Returning to the United States, the shy, self-effacing brothers were the guests of President Taft at the White House, where he awarded them gold medals from the Aero Club of America, later to become the National Aeronautic Association. When they arrived home to lovely weather in June, Dayton hosted the Wright Brothers' Home Celebration, two days of festivities throughout the city that climaxed in a parade and the presentation to the brothers of gold medals from the U.S. Congress, the Ohio legislature, and the city of Dayton. Wilbur and Orville did not let the medals go to their heads. Immediately after the fete, they began their flights at Fort Myer. Their new companies in France and Germany gained business, and in October Wilbur flew a circuit from New York's Governor's Island to the tombs of Ulysses and Julia Dent Grant as part of that city's Hudson-Fulton Celebration, showing the brothers' invention to millions of onlookers. No longer anonymous bicycle mechanics secretly conducting aviation experiments on the secluded North Carolina coast, Wilbur and Orville Wright were celebrities in Europe and North America.

In the autumn of 1909, building on the popular excitement that arose from Wilbur Wright's flights at New York's Hudson-Fulton Celebration in September and October, the brothers decided that the time was right for establishing a U.S. Wright company. While in New York, after his flights around the city, Wilbur Wright was approached by Clinton R. Peterkin (1883–1944), a twenty-six-year-old former office boy for J. P. Morgan and Company. Viewed as "mysterious" by the press, Peterkin was the son of John A. Peterkin, who operated a livery stable in Brooklyn before his 1895 death, and his wife, Helena Rhodes, who became active in religious charities in the decades after her husband's death. Clinton Peterkin never found the success and fortune of Morgan, working in real estate in Florida during the 1920s and 1930s and then in sales in California before he died. Amid the Great Depression's poor real estate market, he felt forced to write to Orville Wright in search of a loan of a few hundred dollars. But in 1909 Clinton Peterkin was young and full of energy and eager to have some role in the new world of aviation. Peterkin proposed that he work for the Wrights, using his connections with the city's business community to create a company to build and sell Wright airplanes. Wilbur Wright was cordial with the young would-be facilitator. He expressed interest in Peterkin's ideas but told him that he and his brother were choosy entrepreneurs. They would only be involved in a corporation in which "men of consequence" with "names that carried weight" invested.[1] Peterkin left the meeting energized. Over the next few months, he responded to this challenge by assembling a group of prominent, wealthy transportation and steel executives and lawyers as investors. Coded telegrams sped from New York to Dayton, with Peterkin keeping the brothers informed on his progress of assembling investors, especially as he and lawyer De Lancey Nicoll—who would be one of the lawyers representing Morgan when he testified before the House of Representatives' Pujo Committee on the money trust in 1912—prepared to submit articles of incorporation to New York's secretary of state.

Satisfied with Peterkin's work and pleased that their names would be associated with "men of consequence" who would enable the company to start with a strong financial foundation, the Wrights traveled to New York City to meet with their investors and file the articles of incorporation on 22 November. After the incorporation of the Wright Company, backed with an unlisted capital stock issue of $1 million, the Wrights were no longer sole proprietors or part of Dayton's petite bourgeoisie. They were truly capitalists. The brothers had not developed their airplane in isolation, having relied on the work of other aeronautical developers, and they had long maintained a generally cordial relationship with engineer and aviation pioneer Octave Chanute (1832–1910), in Chicago. Breaking the news of the new enterprise to the elderly Chanute, Wilbur wrote,

> We have closed out our American business to the Wright Company, of which the stockholders are Messrs. C. Vanderbilt, [Robert] Collier, [August] Belmont [, Jr.], [Russell] Alger [, Jr.], [Edward J.] Berwind, [Allan A.] Ryan, [Howard] Gould, [Theodore P.] Shonts, [Andrew] Freedman, Nicol [sic], & [Morton] Plant. We received a very satisfactory cash payment, forty percent of the stock, and are to receive a royalty on every machine built, in addition. The general supervision of the business will be in our hands though a general manager will be secured to directly have charge. We will devote most of our time to experimental work.

The Wrights confided to Chanute that they looked forward to returning to their workbenches and engineering careers, confident that a professionally managed company would market their airplanes, defend their patents, and bring them a comfortable income.[2]

Peterkin's connections proved viable. Even as the son of a livery stable owner, a young man who was not part of the same social circles as the barons

FIGURE 2.1. Publisher, aviation enthusiast, and Wright Company director Robert Collier. *Courtesy of the Library of Congress*

of Wall Street, he assembled a powerful, extremely wealthy group willing to invest in the Wright Company and lend their names to the birth of the aviation industry. These private investors included inventor, engineer, and army officer Cornelius Vanderbilt III of the famed Vanderbilt family; *Collier's Weekly* publisher Robert J. Collier, who later became president of the Aero Club of America; Nicoll, a former district attorney for New York County (Manhattan) who became corporation counsel; and August Belmont, Jr., the founder of New York's Interborough Rapid Transit Corporation (IRT, or Inter-Met), which opened the city's first subway line in 1904, and a horse enthusiast (the Belmont Stakes, the third and final leg of horse racing's Triple Crown, is named for his father). Belmont was also a distant cousin of John Rodgers and Calbraith Perry Rodgers, both pioneers in aviation; their paternal great-grandfather, Commodore Matthew C. Perry, was Belmont's maternal grandfather. Andrew Freedman, who, with Orville Wright, served as co–vice president, became rich by investing in real estate. Interestingly, several Wright Company directors—Freedman, Shonts, Vanderbilt, Nicoll, and Plant—also served as directors or officers of the IRT. The duplication did not escape public notice. The *New York Sun*, reflecting on the overlap in the two company's officers, wryly quipped that "the suggestion of a merger [of the Wright Company] with Inter-Met was unofficial, unfounded, and unwarranted," and Wright Company records contain no references to its executives' transit activities in New York City. The men who served as Wright Company directors served on many other corporate boards and knew how to compartmentalize their affairs.[3]

Of course, these men invested more than their names. They opened their bank accounts to help create the Wright Company. All the investors save Allan Ryan, the son of transit and tobacco magnate Thomas Fortune Ryan, contributed $20,000 each (nearly $500,000 apiece, in 2010 dollars) to capitalize the company; Ryan contributed $10,000. Peterkin had less luck in getting his former boss to contribute. Though interested in the company's establishment (and given the code name Gold for use in negotiations by telegraph), J. P. Morgan, who in 1909 was spending most of his time traveling and collecting art, declined to invest. Peterkin knew that most Wall Street titans were more interested in money and status than in flight, and couched his proposal to them in promises of high fiscal returns. Berwind and Gould were not particularly interested in aviation. Their investments in the Wright Company were financial gambles on the ability of the Wrights to use their celebrity to make aviation a profitable industry, with their company as its dominant participant, and they did little more than submit proxies, avoiding periodic board meetings (Gould, amidst a bitter divorce, sailed for Europe a day after the company's incorporation and left the board in 1911, spending most of his time in Britain). Allan Ryan, who was president of the Aero Club of America in 1910 and 1911, also left the Wright Company board in 1911. He personally provided nearly $60,000 of the $230,000 it cost to hold the Belmont Park air meet in the fall of 1910 (after which he castigated the Wright Company for suing to prevent the distribution of the meet's prize money). Later described as someone "never willing to take a secondary place in any corporation in which he was invested" and as someone who "wanted to 'sit at the head of the table' or not at all," Ryan organized his own airplane business, the American Nieuport Company, which held the U.S. rights to the French monoplane. Formed—on paper, at least—in October 1911, the firm quickly disappeared from public view and never significantly affected U.S. aviation. Meanwhile, Robert Collier invested for status. While Collier was interested in aviation and also served as president of the Aero Club of America later in the 1910s, when he invested in the Wright Company he looked to associate his name (and that of his popular magazine, which reached nearly a million homes and businesses) with the Wrights. He wanted his Progressive, reformist weekly associated with new ideas and new technology. He later wrote to Orville Wright that his "interest in the Wright Company was purely based on my admiration for your work," not for any particular financial reward. With many of the directors only passively participating in the company, most oversight came from an executive committee composed of president Wilbur Wright, vice presidents Freedman (the committee chairman) and Orville Wright, and members Vanderbilt, Belmont, and Alger. Alpheus F. Barnes, a stenographer and cashier for Nicoll, Anable, Lindsay and Fuller, De Lancey Nicoll's law firm, also sat on the board in his roles as secretary and treasurer. Barnes, "temporarily in charge until a general manager [was] appointed," became the New York face of the Wright Company, with significant power and influence over its operations—and with significant potential to annoy everyone in Dayton.[4]

The company in which these men placed their money defined its scope broadly. It intended to "manufacture, sell, deal in, operate, or otherwise use, at any place or places on the North American Continent and the islands adjacent thereto, machines, ships, or other mechanical contrivances for aerial and land and water navigation of any and every kind and description, and any future improvements and developments of the same," a statement that could encompass the manufacture of automobiles and boats, as well as airplanes. Company officers decided to pursue these activities through a divided corporate structure different from most other period firms that were not railroads or holding companies similar to General Motors. While companies generally placed their corporate headquarters close to their factories, the Wright Company separated its headquarters from its plant by more than six hundred miles. Alpheus Barnes and at least one secretary worked first from the five-year-old, eleven-story Night and Day Bank Building at Fifth Avenue and Forty-Fourth Street in Manhattan, "one of the best constructed and most modern office buildings" in the city, which hosted "the highest class of tenants, including numerous well-known banking houses, railroad companies and estates." Lunch, if Barnes could afford it, could be taken across the street at Delmonico's. An independent office eventually proved unaffordable, and by the end of the company's ownership by Orville Wright in 1915, Barnes worked from Andrew Freedman's 11 Pine Street office in the new Bankers Trust Building in the Financial District, just off Broadway. The company also maintained a New York sales room at an unknown location for a short time.

But Dayton would be home to the company's factory. This geographical arrangement proved unfortunate for the Wright Company. With its president and general managers resident in Dayton and the headquarters (which managed company finances, cut checks, and attempted to supervise factory operations) in New York, communications between the president or general manager and Manhattan were fraught with problems. Unless he visited Dayton, Barnes knew only what officers in Dayton told him—by letter, telegram, or telephone—and the brothers and Dayton-based employees knew just as much about his activities. The Wrights needed to develop relationships built on trust with Barnes and with other executives to create a company that gave its attention to its products and not to internal politics. Most of the investors did not interfere with the Wrights. Few of them ever visited the Dayton factory, and their busy business lives left them little time for active oversight. They became a group principally interested in the fate of their financial investment in the company (as seen in their response to the company's 1914 patent victory). Barnes, though, was intimately connected with the fortunes of the company, and his relationship with the Wrights became combative and untrusting. Barnes and Orville Wright developed decidedly different beliefs on company management, product development, patent enforcement, and marketing—a situation that contributed to the company's inertia while several more nimble competitors surpassed it.[5]

But the competition of Glenn Curtiss, W. Starling Burgess, and other builders remained in the future, and the Wrights remained at home, the company's only executives in Dayton. While they had years of experience as business owners, the brothers' printing and bicycle operations were tiny in comparison with the scope of the Wright Company. They did not distribute stock or have boards of directors, nor did they require the services of a paid manager. Rarely had the brothers even hired employees. Their one long-term hire, Charles E. Taylor (1868–1956), a skilled mechanic, was not cut from managerial cloth. The Wright Company was a different animal than the Wright Cycle Company or Wright and Wright Job Printers. No longer were the brothers building bicycles to sell within the Dayton cycling community—the Wright Company, somewhat rashly, considered its market to be all of North America. And Wilbur and Orville would not be shaping wood at the lathe or stitching fabric on their own sewing machine; the hands of hired employees, not the brothers, would put together the company's products. As Wilbur told Octave Chanute, he and Orville did not intend to manage the company's daily affairs. With no obvious inside candidate for a managerial position, the Wrights turned to the company's executive board for advice. Surely such wealthy men with experience in managing large companies could find someone who would ensure that the Wright Company would prosper? The brothers would have to work with their new investors to find out.

As they began to exercise their new offices, the brothers especially relied on the advice of Russell A. Alger, Jr., of Detroit, the only Wright Company director aside from the brothers based outside New York. Alger, an executive with the Packard Motor Car Company, which built luxury automobiles, came from a privileged background. He was the son of Russell A. Alger (1836–1907), a former Michigan governor, U.S. senator, and secretary of war under President William McKinley and was a leader in the effort to move Packard from its Ohio origins to Detroit. Aside from Packard, he was a corporate officer in more than a dozen other concerns, and "one of the few 'born millionaires' who have made good." First making acquaintance with the brothers in 1908, when he wrote to a convalescing Orville after his Fort Myer accident, Alger quickly became a frequent correspondent of the Wrights. He arranged a purchase of an airplane from the brothers in the summer of 1909 and spent part of the summer refining his balancing skills by riding a motorcycle at top speed (sixty miles an hour) through Detroit, being trailed by his chauffer. That summer, Wilbur and Orville visited him at his office in Detroit as they considered their commercial options. Alger spent much of the summer trying to arrange business ventures with the brothers, going so far as to arrange meetings with other businessmen—meetings that he canceled when the busy brothers neglected to reply to his letters in time. Alger even recommended that his patent attorney, Edward Rector, work with the Wrights' lawyer Henry Toulmin. While not one of the financiers contacted by Peterkin—Alger and his younger brother Frederick made a separate, less well-funded proposal to the brothers in which each Alger would own one-quarter of a company capitalized at $100,000, while each Wright brother possessed 17 percent, with the remainder in the company treasury—they invited him to invest on the same terms as those in New York and prom-ised him a seat on the Wright Company's executive committee. They also told him that they would "be given a rather free hand in selecting the active managers of the concern." Alger, thrilled to be a part of the business, threw himself into its affairs. He looked for a good deal on crankshafts for the airplanes (he believed he could get them at a price $12 apiece less than what Curtiss paid, through one of his other companies), and asked if the broth-ers wanted him to pursue a lead from one of his plant superintendents on a Bethlehem Steel employee who was "the ablest manager of a machine shop he [the superintendent] ever ran across." The brothers were not interested in that suggestion, but they turned with interest to a more personal, even nepo-tistic recommendation, one that proved fateful for determining the Wrights' ongoing relationship with company operations. Alger proposed to Wilbur and Orville that the company hire his nephew, Frank H. Russell, to come to Dayton to be its general manager.[6]

Why Frank Russell? Aside from his connection with the younger Russell Alger, he did not seem an obvious choice. The new general manager was a thirty-two-year-old Yale graduate and yachting enthusiast who was then the president of the Automatic Hook and Eye Company, which made hookless fasteners, in Hoboken, New Jersey, and his only aviation experience was as an interested observer.[7] Though he had relied on family connections through Alger or through his wife's relatives for his previous jobs, Russell had experi-ence in running a big business. Before joining the fastener industry (the mod-ern zipper would not be invented until 1917), he spent five years in Quebec as both a sales manager for Russell Alger's Laurentide Pulp Company and as a U.S. consular agent in the town of Grand-Mère. Among his compatriots in Yale's class of 1900 were artist Frederick Remington; George Whipple, one of the 1934 Nobel laureates in medicine; and future U.S. senator Joseph M. Mc-Cormick—though these connections would not benefit him in Ohio. Only five years younger than his uncle, Frank Russell was slightly acquainted with the Wrights, having met them in July 1909 in Virginia, where he "had a fine time with them" during one day of the trials of the Signal Corps's airplane at Fort Myer; he and Russell Alger also encountered Wilbur Wright at Governor's Island during the Hudson-Fulton Celebration that September. Alger told the brothers that Russell "has had a great deal of factory experience under very adverse conditions and is fully qualified in that direction. He has had also a good deal of business experience. He is a natural mechanic." Addressing the brothers' concerns of nepotism, he noted that he did "not think his being a relative of mine should prejudice him in your mind, although it naturally does prejudice me a little against him and makes me doubly careful in recommending him to you."

Valuing the opinion of someone as successful in business as Russell Alger, and valuing their personal relations, Wilbur Wright agreed to meet with Frank Russell in New York in early December. Russell, whose company in Hoboken was struggling, was looking for new opportunities and eager to meet Wright. There, in the luxurious, baroque Park Avenue Hotel on Thirty-Second Street, they discussed the brothers' plans for the company, which Wright hoped would build at least twenty-five airplanes during the next year. He told Russell that as manager he would have "general oversight of details of construction," sales, and contracts and offered him a salary of $3,000. Russell, excited about the new opportunity, agreed to Wright's proposal, and the board formally hired him at the end of the month with duties of "general manager under the general direction of the president.... It is hoped that in a short time you will be able to assume full charge of the management of the company's business." In January 1910, Russell, with his young family in tow, reported to Dayton to take up his responsibilities managing a company that was preparing to start production in temporary quarters rented from the Speedwell Motor Car Company.[8]

Though his uncle praised his mechanical skills, Frank Russell was a manager, not a mechanic, and he needed to hire a staff to actually build airplanes. While Charlie Taylor was a skilled mechanic, and soon went to work building engines for the Wright Company, the firm needed other skilled workers to shape wood, sew wings, and assemble aircraft. Before 1909, the Wrights employed almost no one besides Taylor and, occasionally, their elder brother Lorin or sister Katharine for bookkeeping tasks. While Katharine and Lorin would be involved with the company when Orville sold it, in 1915, they were otherwise occupied in 1910 and were not involved in its organization or operations. Nor was either sibling the sort of skilled worker the company wanted. To attract such people, the company advertised in Dayton newspapers. Never having attempted to build airplanes commercially before, the brothers and

Russell were unsure just how many workers they would need to satisfy orders in a timely fashion. They optimistically estimated that the factory would eventually employ eighty people—though Wilbur Wright acknowledged that the novelty of building airplanes commercially made it difficult for the company to obtain workers with the proper skills. In New York in the spring of 1910, he stated that "the great difficulty at present . . . is to get the right kind of hands in our factory. We have to teach each man his particular duty, and that takes a whole lot of time." Even in its earliest days, anticipating a future when eighty people crafted airframes and built engines, the company did not anticipate growing into a huge concern spread across town. No campus like that of NCR, spread across south Dayton, for the Wright Company. In any event, such a campus would not be necessary. By the middle of 1911, at the height of its production, the Wright Company employed nearly sixty people in Dayton, its maximum staff. An offer of employment from the Wright Company was not a guarantee of a long-term job, however. Lack of business caused the staff level to fall to "not more than 20 receiving regular employment" by August 1916. The Wright Company was always a minor employer in Dayton, where the combined staffs of National Cash Register, the Barney and Smith Car Company, and the Dayton Malleable Iron Company numbered into the tens of thousands.[9]

The rapid perfection of the assembly line, best exemplified by Henry Ford's Highland Park factory, in Michigan, and used throughout the twentieth century in Dayton by NCR, was still a few years in the future for automobiles, and even further for the aviation industry, where it did not gain widespread acceptance until the 1920s with Curtiss and Ford. The people the Wright Company hired did not staff assembly lines. Instead, as a result of the company's batch production methods—common in the early airplane industry and in such fields as locomotive construction, the production of machine tools, and jewelry—they worked in what historian Philip Scranton terms "functional clusters." In a factory arranged in such a manner, a product—an airplane, for the Wright Company—moves between workers at different job stations. This production method required versatile skilled workers and had little room for unskilled or semiskilled labor. The scientific management principles developed by Frederick W. Taylor at Bethlehem Steel in 1898 and 1899 were of no use in a factory (and industry) where specialized craftsmanship and building individual airplanes to the dictates of a specific order reigned supreme. The Wright Company was more reflective of the American system of manufacturing that grew with the development of interchangeable parts and jigs during the nineteenth century; its production was more akin to the building of a McCormick reaper than a Model T. And Taylorism was useless in a factory where workers produced only a few of the company's signature products each month, as reducing a worker's motions to only those absolutely necessary to speed production was irrelevant.[10]

Speed measured to the second was not something the company considered in creating workspaces. Wright Company production methods and employee demographics are evident in a series of photographs taken throughout the factory in 1911. These photographs, likely commissioned by the company for publicity purposes, given the brief descriptions on some of the prints, provide one of the best records of the race and gender of the company's employees (though who actually took the photographs or how the company actually used them is not clear; they do not appear in period trade journals or company catalogues) and of the factory's interior arrangement. Existing company records are generally silent about labor. There is no archived roster of employees, and contemporary city directories, which generally included a person's profession as part of his or her listing, rarely note the Wright Company as a person's employer.[11] Still, the photographs give a general impression of how the Wright Company was staffed and who answered its newspaper ads. During the early twentieth century, factory employment in transportation-related industries such as automobile and railway car manufacturing was the province of men; men working in urban factories were usually white, with jobs requiring lesser skill often held by immigrants from southern and eastern Europe. All available records indicate that the Wright Company employed native-born Americans, not immigrants, and its small staff was predominantly male and white. The 1911 photographs show at least a dozen individuals—all Caucasian—going about their jobs. Existing company records, city directories, and photographs provide no evidence that the Wright Company ever employed (or ever made any effort to employ) African Americans, even as janitors. Dayton was not an advantageous city for African Americans looking for factory work; blacks looking for industrial jobs in the city were generally limited to working in small plants specifically created by white capitalists and philanthropists to employ African Americans.[12]

The photographer, who had the foresight to record the names of some of the people he photographed, caught well-dressed men working the lathes, assembling motors, painting airplane parts, cutting wire, and undertaking other production tasks. The men appearing in the photographs moved between workstations, suggesting that workers on the factory floor assisted in a variety

FIGURE 2.2. General assembly department, 1911. The unidentified man on the left is working on a Model B (tail number: 2), likely for the exhibition department. *Courtesy of Wright State University*

of areas besides their particular specialty. One worker, Frank M. Quinn, appears in the photographs showing the factory's wire-cutting, woodworking, and radiator departments. Quinn, like most of his colleagues, was an Ohioan, for people did not travel across the United States or the Atlantic Ocean to work in the Wright Company's factory. Census records show that Quinn and the other identified men came from Ohio and surrounding states and were of northern or western European extraction. They answered advertisements the company placed in newspapers for a "first-class machinist," a "first-class wood shaper—hand," or "first-class wood workers." Draftsman Louis Luneke, whose mother emigrated from Germany, was born in Ohio, while both of painter Harry Harold's parents came from Ohio and woodworker John Steinway's were Kentucky natives. Reliance on local, native-born workers typified the early aviation industry's factories. Workers at the Curtiss factory in Hammondsport, New York, also came from the village and its surroundings; only with its expansion to Buffalo as the First World War commenced did Glenn Curtiss employ significant numbers of individuals from outside New York or from countries other than those in northern or western Europe.

Meanwhile, in Marblehead, Massachusetts, a town of 7,300, the awkwardly named Burgess Company and Curtis Inc. also drew its employees from native-born stock. Even the largest immigrant community in the historic maritime town was North American—4.5 percent of the town's residents hailed from Canada, with the largest European community coming from Ireland (3 percent). Marblehead had no significant population of southern or eastern European origin.[13]

In 1914, Orville Wright marched with his father and sister in a parade in Dayton supporting women's suffrage, but his company's division of labor was less progressive. It was, however, representative of the time. The roles of the women on the Wright Company's staff reflected the mores of early twentieth-century U.S. industry as a whole and of Dayton specifically, where there were few women in positions of influence in business or industry and, unlike in Cleveland and Cincinnati, few leading Progressive reform campaigns. The small number of women the Wright Company employed worked at traditionally female tasks, providing clerical and sewing services. At least two women—Maude A. Thomas, listed in the city directory as a secretary and bookkeeper, and Augusta E. Smith, listed as a clerk, worked in the wood-paneled front office, while Ida Holdgreve worked in a corner of the factory sewing linen wing coverings. While Smith's path to the Wright Company is unclear, Thomas, one of the company's first employees, came to its employ through Frank Russell, who hired her in January 1910 from Dayton's Jacobs Business College. Holdgreve, from a rural, working-class background in northwestern Ohio, had no business college training. She responded to a newspaper advertisement searching for someone to do "plain sewing" for the company. In addition to sewing wing coverings for new airplanes, she also mended the wings of airplanes damaged in use by aviation school students and by the company's exhibition aviators. Though Holdgreve worked on some of the first airplanes to be sold commercially in the United States, she waited nearly sixty years to actually ride in one, enjoying a ten-minute flight over Dayton and the former Wright Company factory in 1969. "The clouds looked just like wool," she told a newspaper reporter after returning to the ground. A female machinist or lathe operator, however, was not someone a visitor to the factory floor would see. Only men appeared in photographs of other company departments, and no existing evidence suggests that women worked beside them—or, indeed, attempted to work beside them—in skilled occupations on the factory floor.[14]

CORNER OF SEWING DEPARTMENT

FIGURE 2.3. Ida Holdgreve, the only woman to work on the factory floor, sewing, 1911.
*Courtesy of Special Collections and Archives, Wright State University*

FIGURE 2.4. Unidentified women (perhaps Maude Thomas, Augusta Smith) in the factory's
front office, 1911. *Courtesy of Special Collections and Archives, Wright State University*

Whether Holdgreve, Thomas, Smith, or any other female employee was given pin money or earned a living wage is unclear, given the fragmentary company records, but the former is more likely than the latter, since society expected that younger, white women would work only until they married. But no Wright Company timecards exist today. Remaining records suggest that only general manager Frank Russell and exhibition department manager A. Roy Knabenshue drew salaries, set by the directors in New York. Other managers—whether Russell or the Wrights in Dayton, Alpheus Barnes in New York, or some combination of them—set the pay rates for Dayton employees. Pay on the factory floor in 1913—"top wages" among Dayton employers, according to machinist Tom Russell—ranged from 25 to 37.5 cents per hour for a sixty-hour workweek (reduced to a forty-eight-hour week by 1915, as production levels fell), depending on a worker's experience level. Indeed, the Wright Company, with few unskilled or semiskilled positions when compared with Dayton's behemoth companies, paid relatively well when compared with them, though like many other employers it provided no pension or health insurance. The minimum wage for workers at National Cash Register, which provided a pension and maintained an extensive employee welfare program that supplemented its wages, was $9 per week during the 1910s (between 18.75 and 22.5 cents per hour). NCR and Barney and Smith both provided many more jobs for unskilled and semiskilled workers. Barney and Smith relied on a "colony" of immigrant Hungarians for most of its unskilled and semiskilled labor, paying the workers—who lived in a fenced area of Dayton known as the Kossuth Colony—approximately $9 per week for a seventy-two-hour workweek.[15] Barney and Smith also paid its Hungarian workers partly in scrip, redeemable only at the colony's North Dayton Store Company (the Wright Company always paid cash wages). Wages for skilled craftsman at Barney and Smith, who were not typically Hungarian, were partly based on piecework and bonuses; their base wage rates are not extant. Dayton was an industrial city, and the large concentration of working-class individuals employed at these companies and in the city's many other factories created an environment that would seem to engender labor activism and union activity. But, even as unions battled for recognition across the United States, Dayton—and the Wright Company factory—remained quiet.[16]

Throughout their lives, the brothers, who were usually quiet about most public issues, kept their opinions of labor unions to themselves, though the Republican Party, which they and their family usually supported, was not especially friendly toward labor. Neither brother ever had a reason to join a union. Their company's workers, however, might have been more amenable to joining. From the New York shirtwaist strike of 1909 to the Ludlow Massacre, in Colorado in 1913, the first decades of the twentieth century were years of conflict for the U.S. labor movement. Workers attempted to organize both by craft (unions connected with the American Federation of Labor, AFL) and by class (the Industrial Workers of the World, IWW, which asserted that craft unionism divided workers' combined strength). The Socialist Party of America, on whose ticket Eugene Debs earned 6 percent of the nationwide popular vote for president in 1912, maintained an active local presence in Dayton. However, the violence between workers and management that left dozens dead in the Colorado coal fields did not rise in the Miami Valley. Indeed, Dayton was not a unionized city but one dominated by the open shop. A dispute in 1901 between unionized NCR workers and management over the actions of a company foreman led to the establishment of the open shop at NCR and at most other Dayton manufacturers, many of which (though not NCR) joined together as the Dayton Employers' Association under the leadership of John Kirby, Jr., the general manager of the Dayton Manufacturing Company, a fabricator of metals and of hardware for railway cars, to vigorously protect the open shop. The Dayton Employers' Association was one of the first such organizations created in the United States.[17]

Tom Crouch writes that John Kirby was a friend of the Wrights who, with Wilfred Ohmer of the Recording and Computing Machines Company, recommended in 1903 that the brothers employ Henry A. Toulmin as their patent attorney. But the friendship with Kirby seems to have gone only so far. Kirby never dined with the Wrights, nor did they maintain a correspondence with him after they became famous, and his memoir does not mention them. Valuing at least a minimal presence in the local business community, the Wrights personally paid the company's dues to belong to the Dayton Chamber of Commerce, which NCR president John Patterson established in 1907, but there is no direct evidence that the Wright Company ever joined the Dayton Employers' Association or its sister organization, the National Association of Manufacturers. It did not need to. Aside from paying wages that would make workers less likely to feel aggrieved, the Wright Company employed too few people in individual crafts to make worker representation under the model of the American Federation of Labor feasible. Two or three men did not constitute a United Brotherhood of Carpenters local. Furthermore, the company's workers were not the sorts of men likely to join the IWW: they were not unskilled or semiskilled, and they were rooted to Dayton, not migratory laborers. Neither interviews of company workers nor extant company records mention organized labor, in either a favorable or an unfavorable light. The Wright Company faced many problems during its years in Dayton, but fighting organized labor was not one of them. Indeed, unions found organizing early airframe workers difficult. The AFL's attempts to organize workers in the industry by craft in the 1920s went nowhere, and the International Association of Machinists did not attempt to organize the industry until 1935.[18]

The situation in Hammondsport was little different. In his study of Curtiss workers, historian John Olszowka notes that wage earners across the industry "remained outside the labor movement throughout the first three decades of the industry's existence" and that "the independence, the small shop environs, and the personal contact with employers fostered the creation of a worker ideology that placed an emphasis on promoting the advancement of aviation" and not on class conflict. The small size of the Wright Company throughout its existence supports this concept. Its limited payroll—especially when compared with NCR or Barney and Smith—and relatively small factory likely contributed to a more cooperative ethos between the skilled craftsmen on the factory floor and the company's managers and officers who had the opportunity to have personal contacts with each employee. Still, such personal contacts went only so far. Machinist Fred Kreusch noted that the Wrights ate their lunches at home, not at their office or at the factory. Kreusch also noted that neither brother seemed to have much of a social life nor, unlike Charles Kettering of NCR and General Motors, for whom Kreusch later worked, enjoyed joking with workers while on the factory floor. But the brothers were not always the serious, unsmiling characters they appeared to be to company employees. Both were beloved uncles, and Orville, in particular, was renowned among his nieces and nephews as a practical joker, "a terrible tease" to those he knew well, according to his niece Ivonette, Lorin's daughter. One commonly recounted story, told by Ivonette's brother Horace, known within the Wright family as Bus, had Orville playing with food at a Sunday dinner. Bus liked mashed potatoes, and Orville noted that his nephew's plate always seemed to be heading toward the dish of potatoes, while surreptitiously pulling the boy's plate toward the serving bowl by an attached string. But Orville the puller of his nephew's food did not walk the factory floor in west Dayton; Orville the vice president or president did, and practical jokes were inappropriate around machinery. But when the brothers walked into the New York meeting that created the company in November 1909, the concrete for the factory floor had not yet been poured. Though Dayton was an industrial city, proud of its many factories, the Wrights wanted a brand-new facility for their company and their workers to call home.[19]

# 3

## "A Substantial, Commodious, Thoroughly Modern Factory"

*The Wright Company Enters the Market*

Aviation was a new industry, in Dayton and in the United States. The 1910 *Statistical Abstract of the United States,* issued by the federal Department of Commerce and Labor, unsurprisingly ignored the new field (which had no statistics worth recording). If Wright Company employees were going to build a substantial number of airplanes, they needed an appropriate facility in which to do so. With a wingspan of thirty-nine feet, the Model B, the company's principal production aircraft of 1910, required a large, open area for ease of assembly. Other aircraft companies would repurpose existing industrial buildings.

Glenn Curtiss used his motorcycle factory, where he also continued to build motorcycles until 1912, and Starling Burgess his shipyard, but the Wrights could not easily adapt their former bicycle shop into a factory. The 1127 West Third Street building in Dayton, a block from their home, continued to suffice for their personal office, but the structure's largest room could barely accommodate one airplane. The Wright Company still found the building useful for the brothers' offices and for some work on airplane motors or experiments within its walls, and the brothers continued to pay Charles Webbert rent until 1916. But the brothers were not worried about finding a place for their new enterprise. Plenty of land was available around Dayton, and the company decided to build an entirely new factory on a new site. The Wrights began to survey their possibilities.

While the brothers made no serious inquiries about factory sites away from their hometown, local boosters were initially concerned that the Wright Company, after placing its headquarters in New York City, would site the factory somewhere other than Dayton. The *Dayton Daily News* expressed that local angst, telling the community that "there is a sentimental reason for [the factory] being here, but in industry sentiment does not always overcome physical advantages." It urged Daytonians to "do something more than sit down and wait" for the Wrights, for "it would be a crime, almost, for the aeroplane factory to be located elsewhere." The Dayton Chamber of Commerce, wanting the city to be the birthplace of the aviation industry as well as the birthplace of aviation itself, took up the *Daily News's* challenge and worked with the company to find a site in the Miami Valley. A wire story reprinted in the *New York Times* and several other papers made a rather wild claim, asserting that the Wrights had purchased more than seven hundred acres for a factory near Tippecanoe City, in Miami County, north of Dayton on the Dayton and Troy Electric Railway line, in September 1909. At the time, Orville was flying in Germany and Wilbur was busy preparing for the Hudson-Fulton Celebration flights; the brothers were not even on the same continent. Wilbur quickly put this rumor to rest, at least locally, telling the *Dayton Daily News* and the *Dayton Herald* that the story was rubbish, "another phase of the false rumor which has been circulating for about a month regarding us." The wire services did not pick up Wilbur's denunciation, and the rumors spread as far as Minnesota and Virginia. Other stories had the Wrights buying land near the city of Springfield, northeast of Dayton, or deciding to build their factory at their Huffman Prairie testing field. None of these reports were accurate. The brothers insisted that the factory be built within the limits of the city of Dayton.[1]

Wilbur and Orville had long considered their options in the city. In 1909 they purchased several lots at the corner of West Third and Broadway Streets, eventually using the land for a commercial and apartment block, the Boyd Building, which opened after Wilbur's death. Their factory was also destined for a site within city limits. In quashing the idea that he would agree to build airplanes in Miami County, Wilbur asserted, "we will go right ahead building our machines here in Dayton." Initially, the brothers wanted to stay in their home neighborhood, where they had operated their printing and bicycle repair business since 1889, but they could not find an appropriate location. Instead, the company purchased approximately two and a half acres of land a little more than a mile further west, off Third Street, near the Central Branch of the National Home for Disabled Volunteer Soldiers. The property was a quick five-cent streetcar trip from the brothers' home and office on the City Railway Company's line. Dayton's population grew by more than 35 percent between 1900 and 1910, and the city grew with it, creating new wards on land previously unassigned to one, including the company's new property (which was in the Sixth Ward). This part of west Dayton was not then a hive of industry but "a residence district where a large proportion of the people own and take pride in their own homes." The Wright Company, which had no particularly special technical requirements for its factory, would be building among homes, on eighteen separate lots that it bought for a total of $4,950 from grocer Lewis Clemmer and his aunt Sarah Clemmer Bish, a homemaker who was the wife of a deliveryman. The Clemmer and Bish families were long established in western Montgomery County, and Dayton's voters elected Sarah's son George W. Bish to the office of city auditor in 1909. Their land was platted, with streets mapped out, but it was still in agricultural use and planted in corn when the company acquired it, in August 1910, and the company needed to work with the city to make the site ready for a factory. Through city engineer F. F. Celarius, the Wrights and Frank Russell worked quickly with the city to vacate two alleys crossing the site that interfered with the proposed building location. With the site ready, workers from the John Rouzer Company, one of Dayton's major construction firms, began construction of the first factory building as summer turned into autumn.[2]

But the Wright Company's staff still had airplanes to build during 1910. While the Wrights, Russell, and the board went through the process of finding and purchasing land, consulting with an architect, and engaging a builder, Dayton industrialist Pierce Schenck provided the Wright Company's carpenters, painters, and metalworkers a provisional home in a 7,035-square-foot

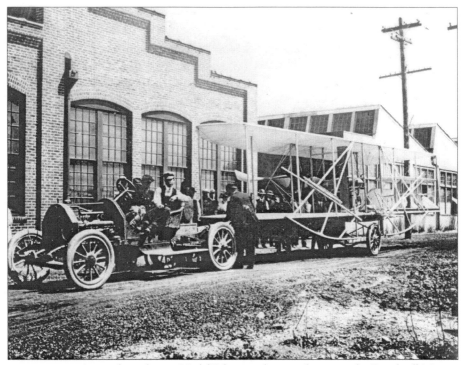

FIGURE 3.1. A crowd watching a Model B leaving the rented space at the Speedwell Motor Car Company factory, 1910. *Courtesy of Special Collections and Archives, Wright State University*

temporary building at his Speedwell Motor Car Company factory on Essex Avenue (now Wisconsin Boulevard), in Dayton's Edgemont neighborhood. Speedwell, which produced automobiles in Dayton from 1907 to 1914, rented the Wright Company the space from January to November 1910 for $250. Schenck, already prominent in Dayton as an executive of Dayton Malleable Iron and a bit of an inventor himself (accumulating six patents over his life), and his secretary, Henry Stoddard, a man "who probably had a wider acquaintance among the mechanics of Dayton than any other person," showed great interest in the commercialization of aviation, with Stoddard announcing his pride in having "the first plant devoted" to the "manufacture of aeroplanes in America" associated with his company. Noting that airplanes and automobiles appealed to different customers, Stoddard stated that his company had "no fear whatever" that providing manufacturing space to the airplane maker marked "the beginning of serious competition between the motor car and the craft of the air." Rather, Stoddard had great expectations for his tenant. He optimistically believed that there would be demand for at least a thousand Wright airplanes in 1910, though how the company was supposed to build so many in the space he made available he did not discuss. While he was happy to help the Wright Company establish itself, Stoddard did not try to keep it based on Essex Avenue. Speedwell wanted to use the land it rented to the Wright Company for its own expansion, while the Wright Company found the space it leased to be too small for the company's workers and machinery to be productive. Intending to expand its own factory and its production of touring cars, Speedwell razed the temporary structure after the Wright Company moved out. However, the Great Dayton Flood of 1913 heavily damaged the Speedwell factory complex and drove the automobile maker into receivership. The site later became part of a General Motors automobile plant, but a century after it hosted the Wright Company, it was a vacant, concrete-covered expanse.[3]

After less than a year at the Edgemont site, the Wright Company bade farewell to Speedwell and began to transfer operations to its new building in west Dayton on 6 November 1910, when general manager Frank Russell moved his office—though it took another few weeks for workers to connect the building to Dayton's electrical grid and completely relocate assembly functions. The company's new home was a one-story steel-frame building designed by Dayton architect William Earl Russ with approximately 13,800 square feet of available space on one floor. The new structure gave the company nearly twice the space it had had in its shop at Speedwell. The thirty-two-year-old Russ, who designed Dayton's Memorial Hall and who later became a prominent architect in Indianapolis, created blueprints for a functional building with little ornamentation. Though the structure had large windows and a skylight running its entire length, as well as interior walls painted white to reflect available sunlight, it did not have multiple floors, large steel-framed windows, or reinforced concrete. Stylistically it was a variant of the daylight factory made famous by Ford Motor Company's plant in Highland Park, Michigan (1910), designed by Albert Kahn, and of several contemporary buildings in Buffalo, New York. The construction of a building connected with the romance and novelty of flight attracted significant local attention. The *Dayton Daily News* reported that the "crowds which gather every Sunday afternoon seems [*sic*] to indicate some astonishing aerial performance, but the building is the usual factory structure and will not fly away." A Chicago journalist, responding to the factory's supposed novelty, wrote that the plant, regardless of its connections to aviation, was an ordinary place, "as businesslike and commonplace as a shoe factory." Crowds soon learned that the activities within the building provided little entertainment; the Dayton press never again wrote of large groups gathering at its doors, even after the facility opened, in November.[4]

Early aviation in Ohio and other continental climes was a seasonal activity. Pilots of the Wright Company's airplanes sat on the lower wing, completely exposed to the weather, and few flew frequently in the Ohio winter. But the company intended for airplane building to occur throughout the year. Photographs taken in 1911 show radiators hugging the building's interior walls, but the new factory, with its large doors, many windows, and lack of insulation, was a chilly workplace when it opened, in late autumn 1910. The frugal Wrights, who as young printers had built a printing press out of scavenged parts that included a buggy top and an old gravestone instead of buying a commercial model, told Russell Alger that a heating system would be a later installation. A million-dollar capitalization, not being liquid cash, went only so far. "We will probably use stoves for the first winter," they wrote, a winter that they mostly devoted to the construction and maintenance of airplanes for the exhibition department, as few outside orders were arriving. Though well capitalized, the company was young and its managers viewed centrally heating the large, open, high-ceiling factory as an unaffordable extravagance that year. With architectural fees added to the cost of construction, the total bill for erecting the factory came to about $13,000 (roughly $237,000 in 2010). The company described its home in Dayton as "a substantial, commodious, thoroughly modern factory . . . equipped with the best tools obtainable." According to the Chicago Tribune's Irwin Ellis, it was a monument to the success of aviation. In each of the factory's three divisions (motor shop, woodworking, assembly), Ellis wrote, Wright Company workers took special care in the various steps of constructing their vehicles, for "the aeroplane has been regarded until recently as too dangerous for use by any but the most reckless sportsman. . . . It is only by turning out the strongest machines possible that the aeroplane makers can show that the flying machine has reached a point where it is safe when used with caution and discretion as to proper flying conditions." To ensure that its workers could accomplish that task, the Wright Company purchased much of its industrial machinery from the J. A. Fay and Egan Company of Cincinnati. The company installed electric lights throughout the work areas, but leather belts powered by a small, onsite coal power plant drove the lathes and drills that workers used.[5]

The Wright Company now had its own dedicated factory and was ready to build airplanes without having to consider the views of a landlord. It just needed orders. Though production space doubled in the move from

FIGURE 3.2. The Wright Company factory in late November or December 1911. *Courtesy of the Library of Congress*

Edgemont, growth in production capacity is more difficult to determine. Estimates vary on just how many "strong machines" the company's employees, working in a pre–assembly line era, could produce in a specific time period in the 1910 building. Wright biographer Fred Howard has suggested that workers could build a maximum of two airplanes each month while over at the Speedwell site, and four per month in the new facility; historian Tom Crouch places the new building's maximum production at "two airplanes a month, complete with engines." William H. Conover, who worked for the Wright Company from 1909 to 1915, recalled building "maybe eight or ten in Edgemont" in 1910, all for the exhibition department, demonstrating the limited outside demand for company products and aviation's niche status in the contemporary American economy. The Wrights themselves were more realistic about their prospects than was Speedwell's Henry Stoddard and estimated that the new building would "be sufficient for an output of 50 machines per year" or approximately one per week, and Lt. Frank Lahm, the army's first certified pilot, who learned to fly through the Wright Company's school, wrote his father in August 1911 that busy workers at the Wright factory were maintaining that level of production when he visited. Regardless of maximum production capacity, the Wright Company did not receive the deluge of orders—a thousand a year—that Stoddard had predicted in 1909 (Stoddard also oddly boasted that, a month after its organization, the Wright Company had actual orders for 175 airplanes), and it would not have been able to fulfill such a high number in any case. Even after 1911, when both buildings were in operation, it could take at least four months from the placement of an order to the delivery of a finished airplane, which directors Vanderbilt, Collier, and Alger discovered when they ordered the first three airplanes that the Wright Company built. Early airplanes were not transported by air, and completed models left the factory as cargo on a rail spur on its north side, with the Southern Railway handling shipments to the southern United States.[6]

The Wright Company may have shipped to the southern part of the country (and the northern, eastern, and western), but it did not ship overseas. Though it was one of the best-capitalized airplane producers in the United States upon its creation, in 1909, it existed in a different industry than did the Wright ventures in Europe. The expansion of the U.S. aviation industry into new products and technologies proceeded more slowly than did Europe's factories. Annual aviation catalogues show how European firms grew more rapidly than the Wright Company and other U.S. builders in the years before the First World War. While Fred T. Jane devoted fifty-four pages of text to U.S. airplane makers in the 1909 first edition of his aviation annual, most of the listed concerns were private individuals building airplanes at home; only two of the nearly eighty monoplane, biplane, triplane, and helicopter builders Jane included—the Wrights and Glenn Curtiss—were capitalized as airplane-building companies. More commercial aircraft building occurred on the east side of the Atlantic Ocean. Jane wrote, "France has been the pioneer of heavier-than-air machines, and at present occupies the premier position in this field." He devoted sixty-six pages to the French industry, which included Léon Levavasseur's Antoinette company and the Farman, Blériot, Esnault-Pelterie, and Voisin concerns, as well as Astra, the French Wright concessionaire, among the ninety different French builders. France's major companies collectively produced hundreds of airplanes annually.[7]

The size of the French industry as compared with what existed in the United States drew some attention in the U.S. press. In the spring of 1909, a New York Times reporter wrote that Paris contained fifteen airplane factories, most of which had opened in the previous six months. Most of the factories in the Paris region, where space was at a premium, with nearly 4.5 million residents, reused existing buildings. Companies found that "almost any disused factory can be utilized as an aviation factory so long as there is enough room

and a good light from the top." While the Wrights were concerned about finding qualified help in Dayton, French firms did not find it difficult to obtain
competent employees. They believed that "with good plans and plenty of room,
any good carpenter could turn out an aeroplane." While, as shown by Jane and
by the number of different models appearing in the Gallic skies, the French
airplane industry was larger than its U.S. cousin, the tools and materials it used
were similar. An unidentified factory visited by the *Times* reporter, a "concert
hall in the rough" in Billancourt, a Paris suburb, was similar in structure to U.S.
factories, with the same sorts of departments, materials, and equipment: a space
for woodworkers to cut the ribs for the wings; space to run wire and assemble
motors; and space to assemble complete airplanes. It clustered its offices by the
building's entrance, with the offices opening to the production floor and its
woodworking and metalworking stations. While the walls of French airplane
factories may have been repurposed, the insides were not. Henri Farman's Aviation Works, which employed one thousand workers by 1914, and Charles and
Gabriel Voisin, whose company became a major producer of aircraft for France
during the First World War, recognized the importance of giving their workers
access to plenty of space and modern tools for making airplanes on a commercial scale. The Voisins even advertised that their company possessed "a real
plant" with "a fully equipped machine shop," suggesting that not every airplane
builder was so well equipped. While the *Times* reporter estimated that the factory he visited could turn out one completed aircraft each week, the collective
capacity of the French industry was much higher than that of its U.S. counter-
part (Blériot aéronautique even employed four different factories to build its
airplanes). By August 1910, French companies had sold nearly three hundred
airplanes to public and private buyers. The U.S. industry did not achieve that
level of production for several more years; the Wright Company never attained
such numbers (nor did Astra, the French company licensed to build Wright
models). In 1915 the U.S. Army Signal Corps found that the Wright Company
could produce only one airplane a week, while Curtiss led the industry, building an average of 2.5 per day. The buildup to the First World War accelerated
developments in Europe; the U.S. industry was late to catch up.[8]

Still, while U.S. builders were aware of developments among European
companies (always protective of their intellectual property, the Wrights considered many of them to be patent infringers), they were not in direct business
competition with them. While airframe patterns crossed the ocean—the contemporary aviation press is full of short notices of people building Blériot- or
Farman-type aircraft from plans—tariffs and the potential of Wright lawsuits

meant that aircrafts actually built by French companies made the transatlantic journey less frequently (often only for meets; Blériot shipped five airplanes to New York for the 1910 Belmont Air Show). And American builders would not have been very competitive in an open market. U.S. commercial production in the first years of the Wright Company lagged behind commercial production in France. Aviation journalist Ernest Jones reported that U.S. builders made nearly 40 percent fewer airplanes in 1911 than French companies had by August 1910 (approximately 1,050). While at least 750 airplanes were built in the United States in 1911, he found that most were built by individuals; only 174 were built by companies (of which 58 were for private individuals, 105 for exhibition pilots, and 11 for government). Jones's figures also show the hesitancy of the U.S. military to acquire airplanes (unlike its European counterparts). As late as August 1913, it owned only six flyable aircraft, while the air corps of France, Germany, and the United Kingdom each flew scores of airplanes and also made significant investments in lighter-than-air aviation. Though U.S. airplane builders needed the U.S. government to be a major customer if the domestic industry was to develop (since few individuals could afford to buy an airplane), between 1908 and 1913 Washington was a poor patron of the field, spending merely $435,000 on aviation. The U.S. outlay was far less than what Germany ($28 million), France ($22 million), or even Chile ($700,000) spent. However, the country where the Wrights first flew did not lack aspirant companies. Many were small, undercapitalized outfits—the May–June 1912 issue of *Aeronautics* listed eighteen new airplane-building or exhibiting companies, most of which were capitalized for less than $50,000 and provided little competition to the Wright Company. Some businesses, though, gave the Wright Company stiff competition from the moment of its creation, and even eventually surpassed the Dayton concern in introducing new types of airplanes and in numbers of airplanes built. The two firms whose histories crossed the most with the Wright Company were Glenn Curtiss's Curtiss Aeroplane Company and W. Starling Burgess's Burgess Company and Curtis.[9]

Curtiss (1878–1930), like the Wrights, had a background in bicycles. Growing up with his mother and grandmother in Hammondsport, New York, Curtiss worked as a bicycle messenger and a photographer before opening his own bicycle shop, in 1900. He soon became interested in motorcycles and in building lighter, better engines for them. Exhibiting a daredevil side that neither Wright possessed, Curtiss also became a prominent motorcycle racer. His light engines became popular among early aviators (though not with the Wrights, who built their own engines for their 1903–5 airplanes with their mechanic Charles Taylor). Through his engine construction for California dirigible pilot Tom Baldwin, Curtiss came to the attention of Alexander Graham Bell, who made him a part of his Aerial Experiment Association. Bell, a renowned scientist and inventor who had extensive experience in patent litigation through his battles with Elisha Gray in the 1870s over the invention of the telephone, would be a key adviser to the Hammondsport inventor in his legal battles with the Wrights. Curtiss designed his famous June Bug airplane for the AEA in 1908 and made his first corporate venture into aviation the next year with Augustus Herring, when the Herring-Curtiss Company took over his existing motor and motorcycle factory in Hammondsport, a village of 1,250 in the wine country around the Finger Lakes, on the shores of Keuka Lake. Curtiss began developing the site in 1904, when he built the first of what became a complex of nearly a dozen factory buildings that surrounded his family's home on a hill above Hammondsport. He expanded operations quickly, employing nearly forty people within a year. Even after he and Herring incorporated, in 1909, his factory still principally produced motors and motorcycles. That year he expected his workers to build one thousand motorcycles, between fifty and one hundred airplane motors, and only eight to ten airplanes. But Curtiss's devotion to motorcycles would not last. His childhood friend and factory manager, Harry Genung, recalled that Curtiss "had a front-row seat in the motorcycle business when aviation came along and pushed the business out the back door." In the manner in which the Wrights relied on their bicycle business to support their aeronautical research, the profitability of the motor and motorcycle wing of the enterprise gave Curtiss some financial stability as he built his airplane company and endured the Wright Company's lawsuits.

Curtiss needed that stability. His business fortunes briefly stagnated as Herring-Curtiss's board engineered bankruptcy proceedings in 1910 to remove Herring from its corporate structure. Herring held 2,015 of the company's 3,600 shares of stock and was supposed to turn over to the firm several aeronautical patents that turned out to be figments of his imagination. Herring quickly sued Curtiss, starting a long legal battle not resolved—in Herring's favor—until 1928, two years after his death. After divesting Herring, Curtiss purchased Herring-Curtiss's corporate assets from the bankruptcy trustee and reorganized the business as the Curtiss Aeroplane Company. Herring-Curtiss, capitalized at $360,000, had attracted some prominent, wealthy investors—including art patron and real estate tycoon Courtlandt Field Bishop, who was president of the Aero Club of America for several years between

1906 and 1910 and invested $21,000, and International Harvester industrialist James M. Deering, a friend of Bishop's, who added $1,000 to the company's books—but it did not have the deep Wall Street connections of the Wright Company. Bishop and Deering lost their investments when Herring-Curtiss entered bankruptcy, and they did not reinvest in Curtiss Aeroplane. Instead, when he reformed the company, Curtiss, hoping to avoid the likes of Herring, looked to Hammondsport and to individuals he knew and trusted for funding. In December 1911, Curtiss and local investors, including local lawyer and judge Monroe Wheeler, Curtiss Exhibition Company manager Jerome Fanciulli, and several other company managers, provided the capital and board members for the formation of the Curtiss Motor Company, which was capitalized at $600,000. Its fortunes would be much brighter than those of its predecessor.[10]

The workforce in Hammondsport remained constant as Herring-Curtiss dissolved and Curtiss Aeroplane began. They may have built airplanes in a partially preexisting factory, but their methods, tools, and machinery differed little from the implements used in Dayton or Billancourt. Curtiss made it easy for interested lay people to learn about his operations, approaching them with a book he wrote in 1912 with Augustus Post, secretary of the Aero Club of America and a friend who happened to own the first automobile registered in New York City. *The Curtiss Aviation Book*'s three hundred pages of biography and public relations contained a detailed description of the Hammondsport factory. Post told readers of a plant with "the latest machines of all types," including a "unique machine" that "carves out propellers from a laminated block of wood." The factory contained individual rooms for different steps in the production process, from brazing to japanning to woodworking to assembly, as well as space for building motorcycle and airplane engines. Curtiss proactively provided factory space for research and development, incorporating a laboratory into the complex. There workers prepared airplane schematics and conducted wind tunnel experiments on new designs. While Wright Company employees used the Wrights' former bicycle shop on West Third Street as an ersatz laboratory, and flight-tested airplane designs at Huffman Prairie (and on the Great Miami River), Russ's design called for no space specifically dedicated to laboratory work. The two companies' differences in labeling workspaces reflected different corporate cultures: one more supportive of innovation (partly as an attempt to avoid patent litigation), in Hammondsport, and one satisfied with the quality and capabilities of its existing product line, in Dayton. As its airplane business grew, Curtiss found itself limited by the resources available to it in tiny Hammondsport, and the company moved its

FIGURE 3.3. Glenn H. Curtiss in his Reims Racer, 1909. *Courtesy of the Library of Congress*

principal factory to Buffalo in late 1914. An elementary school named for Curtiss now covers the site of his first factory.[11]

Sailing, not cycling, was the first business interest of W. Starling Burgess (1878–1947) of Marblehead, Massachusetts. The father of author and illustrator Tasha Tudor, Burgess loved sailing and had rushed from Harvard to join the U.S. Navy during the Spanish-American War. Returning to Marblehead (but not to Harvard), he became a yacht designer and builder, opening his own shipyard in 1905. Three Burgess-built yachts would win the America's Cup regatta during the 1930s. Wilbur Wright's flight during New York's Hudson-Fulton Celebration in 1909 piqued Burgess's interest, and he turned his attention and the work of his shipyard from speed on the sea to speed through the air. After a brief dalliance in early 1910 with Glenn Curtiss's erstwhile partner Augustus Herring (Herring built two airplanes with Burgess), that June, Burgess and businessman and engineer Greely S. Curtis (1871–1947), whose son James served in the Taft administration as an assistant secretary of the treasury, incorporated the Burgess Company and Curtis, capitalized at $80,000. The yacht builder was impressed with the Wright Company's work, and Marblehead and Dayton soon became closely connected. Burgess became one of the first students at the Wright Company's 1911 winter flying school in Augusta, Georgia, learning to fly the Wright Model B; the *Atlanta Constitution* wrote that "he appears to be getting more or less sport out of the lessons." But Burgess's lessons had a more commercial purpose. Between 1911 and 1914, his company built licensed versions of Wright B and C airplanes, which it sold as Burgess-Wright Models F and J. But Burgess gradually fell out of favor with the Wrights. While the contract with the Wright Company bound Burgess to build exact copies of the Wright Company's airplanes, Burgess upset the Wrights by building airplanes that were generally heavier than their Wright Company siblings and were composed of stronger materials. Burgess was the only company that built licensed Wright-model airplanes, paying the Dayton firm a 20 percent royalty on the sales price of each airplane.[12] It also built airplanes under license from Glenn Curtiss and English aviator Claude Grahame-White and designed and constructed its own models. Burgess sold his airplane company to Glenn Curtiss in 1916 and spent most of the rest of his life designing and building yachts. His airplane company's background in shipbuilding and its factory's location, adjacent to water, provided critical testing facilities as the popularity of seaplanes grew during the early 1910s.[13]

Established as a shipyard in 1904, the Burgess Company's principal facility, at Redstone Lane in a residential area of Marblehead, bordered Ladys

FIGURE 3.4. W. Starling Burgess, ca. 1910–15. *Courtesy of the Library of Congress*

Cove in Salem Sound. The town's economy centered on water-based activities such as fishing and shipbuilding; during the early twentieth century shoemaking also employed many locals. Burgess's airplane operation began small, employing only six workers in 1910, but the company's involvement in aviation grew rapidly, employing nearly two hundred by 1916. While Burgess could not provide spacious facilities at this location—lathes and workbenches stood more closely together than in relatively spacious west Dayton—workers used the same sorts of equipment as their colleagues in Dayton, Hammondsport, or Billancourt. Finding skilled workers for the precision woodworking necessary in building early airplanes was not a problem for Burgess, as precision woodworking was a crucial skill in boat construction and was common among Marblehead's working class. Motor making, though, was not a skill in demand in the Marblehead plant. Instead of making its own engines, Burgess purchased the motors for its Wright-licensed airplanes from Dayton (no more than three a month, according to the license), while for its other models it acquired engines from such manufacturers as Curtiss or Renault. Records concerning the company's production capacity when it produced Wright-model aircraft do not survive due to a 1918 fire at its second plant (a facility constructed at Little Harbor, near Falmouth, in 1916 as the company increased production during the First World War). However, the *New York Times* reported in 1916 that the employees working in the factory's twelve buildings could turn out "about thirty aeroplanes a month" (approximately eight airplanes a week), then the second-highest production capacity in the United States after Curtiss's Buffalo factory.[14]

These factories in Dayton, Hammondsport, and Marblehead, which relied on the labor of skilled workers, operated at the end of the craft era. During the early years of the U.S. aviation industry, Henry Ford popularized assembly line production at his automobile plant in Michigan with semiskilled workers repeating the same task at a station on an object moving between stations at a company-set speed. Ford produced tens of thousands of automobiles each year (13,840 Model Ts in 1909; 394,788 in 1915), marketed and priced for sale among the public. Production in the entire U.S. airplane industry until the time of the nation's entry into the First World War never approached such numbers. Unable to keep up with even the U.S. military's desires, the entire industry delivered only 227 airplanes (on 532 orders) to the War Department between 1907 and 1917, and production for the civilian market was also limited. Nor were early airplanes composed of easily replicated and assembled metal components. The wooden frames of early airplanes required the handiwork of skilled lathe operators who could create parts that met the exacting tolerances airplanes required. Former employee Ernest Dubel stated that Wright Company craftsmen made "one or two [of a part] at a time; you didn't make quantities. You couldn't reduce cost much" through bulk production. Nor did airplane formats remain constant. While early automobile models changed little from year to year (the Model T being a prime example), airplane designs changed more frequently; among Curtiss, Burgess, and even the Wright Company, prewar models rarely remained in production for more than a year without undergoing significant modifications. Sufficient commercial demand for airplanes that would necessitate the development of assembly line production in the industry did not exist in the United States until the militarization that occurred just before the country entered the First World War. Only with the advent of fighting in Europe and the probability of large orders from the U.S. War Department did staff levels and the organization of factory labor change significantly at Curtiss (with its move to Buffalo), Burgess (by then owned by Curtiss, with its Little Harbor facility), and other airplane builders. And while militarization brought great profits to airplane makers—and significant losses with the end of the war—the Wright Company found its first years to be its most successful.[15]

# 4

## "Our Machines Are Sold on Their Merits"

*Patents, Profits, and Controversy*

Though neither Wilbur nor Orville Wright would ever have the resources of a Rockefeller or a Vanderbilt, aviation brought them wealth. They made generous Christmas gifts to their brothers and sister, built the Boyd Building in west Dayton as a commercial investment, and built an expansive new mansion, Hawthorn Hill, in the wealthy Dayton suburb of Oakwood, to which Orville moved with his father and sister in 1914. Together they received a $50,000 share of a $100,000 corporate profit as the company's 1910 fiscal year ended, the equivalent of nearly $2.4 million in 2011. During its existence, the Wright

Company's greatest money maker was its short-lived exhibition department, which grossed $271,182.32 from its flights at fairs and competitions around North America in 1910 and 1911 combined. The exhibition department's aviators needed airplanes in good repair, and the factory's workers devoted significant effort to repairing their airplanes and replacing aircraft destroyed in accidents. The military and civil airplane markets seemed wide open to the company, as the initial injunctions issued by the U.S. district court in Buffalo in the Wright Company's patent infringement lawsuit against Herring-Curtiss proved favorable to the Dayton operation. This business environment convinced the company that it needed more factory space to satisfy the expected demand for its products. In the summer of 1911, as Hiram Bingham rediscovered Machu Picchu, high in the Peruvian Andes, and as General Motors became the first automobile company to have its stock listed on the New York Stock Exchange, the Wright Company commenced construction of an 8,450-square-foot brick structure also designed by William Russ, immediately south of the 1910 building that mimicked its style (though, for some unstated reason, Orville Wright found its construction and design "not nearly so good" as the earlier building). The new space—the construction of which company officers and documents generally ignored, suggesting that the project was something about which the Wrights, Frank Russell, Alpheus Barnes, and the other directors agreed—gave the Wright Company greater ability to expand airframe production and build a sufficient number of engines for its own airplanes and for those being built in Marblehead under license by the Burgess Company. The company was ready to grow.[1]

Space, not additional workers or Taylorist scientific management, was what the company needed if it was going to be able to build more airplanes. Historian Philip Scranton notes that for batch producers such as the Wright Company, "expanding output entailed the extension of facilities more than the intensification and restructuring of labor processes." However, the corporate optimism around the new building's construction reflected the company's aspirations, not its actual production. While the exhibition department remained profitable, the Wright factory built only six complete airplanes between January and July 1911, the months immediately before it added more production space. The rather opaque Wright Company sales ledger suggests that Burgess purchased at least five engines from Dayton during that period, implying that the company's engine builders were not constantly busy. Regardless, the company remained bullishly cheerful about its fortunes after its new building opened, boasting that "all indications point to the necessity for

FIGURE 4.1. Interior of the Wright Company factory's building 2 under construction, 1911. *Courtesy of Special Collections and Archives, Wright State University*

a third building during 1912." But the confidence was misplaced. Several fatal accidents, coupled with Orville Wright's aversion to the sensational type of flying the exhibition pilots preferred, resulted in 1911 being the last year the exhibition department flew. Its disbanding deprived the factory of one of its best customers and coincided with the technological stasis of the company's product line. President Wilbur Wright's death, in May 1912, deprived Orville, newly ascended into a corporate office he did not want, of his principal colleague and adviser, and made it even more difficult for a man with limited interest in the day-to-day grind of running a business to build the sorts of airplanes his putative customers wanted. As the U.S. economy entered recession in 1913 and 1914, the Wright Company's fortunes stagnated, and it never added additional production space to the factory.[2]

The Wright Company's torpor was not typical of the industry. Other U.S. airplane builders (including Curtiss and Burgess) grew during the early 1910s, producing more and more airplanes each year, sold at prices similar to those set in the company's conversations between Dayton and New York, but the Wright Company did not. Its stasis arose from several issues. While not a direct cause of the company's eventual decline, the poor national economic situation did not help. For forty-seven of the seventy-two months during which Orville Wright was a company officer, the U.S. economy was in recession, from January 1910 to January 1912, and again from January 1913 to December 1914. Most of the Wrights' attempts to market their airplane took place during years of economic contraction, starting with the Panic of 1907 and ending only with the increases in federal military spending connected with the First World War. But one should not make the overall economy too much of a factor, as both the Burgess and Curtiss companies found ways to build markets and increase sales in the face of recession. Moreover, airplanes were not a mass-produced commodity but a niche product purchased by governments, wealthy individuals, and pilots who hoped to recoup the cost of an airplane by performing exhibition flights at carnivals, fairs, and other public events. The financial wherewithal of airplane purchasers helped insulate the industry from the travails of the overall economy.

Even aviators who learned to fly through the Wright Company's Huffman Prairie school found it difficult to purchase their own airplanes. In 1911, Lt. John Rodgers of the U.S. Navy spent much of the spring in Dayton learning how to fly a Model B biplane, partly under Walter Brookins's instruction. Brookins, a Dayton native, had studied Latin under Katharine Wright at Steele High School downtown. At the end of his training, Rodgers, who had become close to the Wrights during his instruction, wanted to continue to fly. Rodgers was the scion of a distinguished naval family—his father, William Ledyard Rodgers, retired from the service as a vice admiral, and his grandfather, also named John Rodgers, was a rear admiral whose last post was as commander of the U.S. Naval Observatory—but he was not wealthy. He managed to procure half the needed funds from his uncle in Pittsburgh but found it difficult to raise the rest of the purchase price. At the end of April, he traveled to Washington, DC, "to see if he could get some one [sic] to put up a prize for him" so he could afford the purchase. He soon returned to Dayton and was, according to Katharine Wright, "full of schemes." Rodgers approached a friend in the capital, a Mr. Thompson, who agreed to provide $2,500, plus additional funding to pay for maintaining the airplane. But, Thompson arranged a catch—or thought he had. Rodgers's airplane was to be given to the U.S. Navy, and he would only be able to fly it "for prizes etc. etc." when he was off duty. When Thompson and Rodgers met with Capt. Washington Irving Chambers, Rodgers's superior officer, to formalize the deal, he thought that all was well. Chambers had changed, however, and no longer was interested in an arrangement that was "beneath the dignity of the Navy." The "hard-headed" Thompson and the "good

and hot" Chambers entered into a vigorous argument that left Rodgers without an airplane, the latter quickly returning to Dayton to avoid "being involved
in a scrap with the Captain." Rodgers, "a *circus*," according to Katharine Wright,
barely earned $2,500 a year from the navy. Rodgers earned his pilot's certifica-
tion, becoming the second U.S. naval aviator, but he would have to do his flying
on airplanes wholly owned by the fleet.[3]

Rodgers would not have had trouble purchasing the other great transportation technology of the era, the automobile. A 1910 Ford Model T sold for
$780, a price that by 1916 had fallen by more than 50 percent, to $360. While
the automobile industry was becoming an important part of the national
economy (and, through General Motors, of Dayton's economy), the impact
of the early aviation industry was negligible. It occupied a small place in the
economies of the communities in which it was based, unless those communities (like Hammondsport) were very small. The federal government took little
notice of the industry's role in the economy, as aviation did not obtain its own
line in the U.S. Census Bureau's *Census of Manufactures* until 1920. In Dayton, where National Cash Register, the Barney and Smith Car Company, and
Dayton Malleable Iron employed thousands (not the few dozen employed by
the Wright Company), and in Marblehead, which supported the significant
shoemaking, shipbuilding, and fishing industries, aviation remained a minor
contributor to local economies until well after Archduke Franz Ferdinand's
assassination, in Sarajevo in 1914.[4]

The Wright Company principally relied on news articles in the popular
press (which was not always favorable to its interests) and the performances of
exhibition aviators to market its name and its wares among the general public,
limiting its print advertising (see chapter 7) to the aeronautical press. Given
the limited market for airplanes, wider advertising would not have helped the
company work against recessionary times and grow. Even the advertising placed
by the company in the several circulating aviation journals was not a priority
for the Wrights or the board. In a 1912 letter to private-exhibition aviator and
instructor Max Lillie (who died in his Wright B in 1913 during an exhibition
flight in Illinois), board secretary Alpheus Barnes noted, "Our machines are
sold on their merits and therefore we do not conduct a strong advertising campaign nor employ solicitors. We are sure that we build a superior machine and
are greatly gratified by the result of the business done." The previous summer,
general manager Frank Russell, soon to leave Dayton for Marblehead, wrote
in his diary that Orville Wright did "not want to spend any money in Sales
development—Thinks orders will come without effort." Russell also wrote to

vice president Andrew Freedman, "strongly recommend[ing] the appointment of a competent salesman, to . . . take charge of our sales work as well as our publicity department. Up to this time [July 1911] there has been no systematic publication of news items either in trade papers or general publications. Our plant is the only complete aeroplane factory in America, yet the public knows little or nothing of our progress." Russell, who supervised no dedicated marketing or publicity staff in Dayton, believed that the Wrights, Barnes, and Freedman expected that the brothers' name provided enough publicity to build an industry-leading company without undertaking additional work.[5]

Curiously, given Russell's belief that the company needed to employ such a person, the Wright Company did have at least one traveling agent—Thomas P. Jackson—working for it in 1910, promoting the company and its exhibition pilots in Kansas City and beyond, and it also employed André Roosevelt in that capacity in 1911. Russell and Roy Knabenshue needed to be careful with their agents, though. In the spring of 1911, Katharine Wright informed President Wilbur Wright, on business in Germany, that Orville believed that Jackson was embezzling company funds. Simultaneously, Orville informed Wilbur that Jackson, away in New York and in trouble for mismanaging an estate, was unwanted baggage. "He has been in one kind of trouble after another for some months past," wrote the vice president, and he had "told Knabenshue to find someone to take his place." Wilbur remained silent, and Jackson, already a murky and ignored character in the company's extant papers, disappears from the remaining record entirely. In any case, individual promoters could do only so much meeting and greeting of potential clients. The Wright Company recognized the need to market itself more generally within the aeronautical industry and placed half- and quarter-page advertisements in *Fly* and in *Aeronautics* between 1910 and 1915, even during Russell's tenure in Dayton. It also employed the services of the Erickson Advertising Agency, a prominent New York firm, to publicize its products, and sometimes found their work wanting. Orville Wright and Russell, who rarely found agreement on company issues, even agreed that an advertisement proposed by Erickson was "too conservative for the trade Journals." With aviation's specialized market, the Wright Company never advertised in such popular mass-market periodicals as *The World's Work*, *McClure's*, or *The Review of Reviews*, or in local or national newspapers. It did occasionally advertise in the upper-crust *Town and Country*, and during the spring of 1914 Grover Loening wanted to advertise in *Yachting* or *Motor Boating*, believing that "one or two big advertisements in Country Life, Town Topics and the like would do more good than hundreds of small adds [*sic*] in aviation magazines which are seldom ever read anyhow." Loening thought that New York society, where there were plenty of rich aviation aficionados, was a potentially valuable consumer market, and he wanted it to be aware of the company's new Long Island flight school, but the upper-class shelter and society scandal magazines he mentioned never carried Wright Company publicity. Under Frank Russell's management, the company attempted to conduct direct-mail marketing by sending its catalog to members of aeronautical clubs but found it difficult to obtain names and addresses of the members of the many societies across North America. Later, Loening created a list of newspapers and periodicals to receive company publicity, but that effort came too late in his tenure in Dayton and was forgotten after he left. However, his work did not significantly affect the treatment of the Wright Company in the press, which devoted significant column space to coverage of lawsuits over infringements of the 1906 patent.[6]

When perusing the Wright Company's ledger, one is struck by the number of pages—four—specifically listing the payments to lawyers and court costs connected with the defense of the Wrights' 1906 airplane patent. Sales occupy just a half page more space than do legal expenses. Defending the patent, not developing new kinds of airplanes, was the Wright Company's principal activity between 1911 and 1914. The patent battles gained significant coverage in the popular and aeronautical presses, challenging company executives to try to maintain a positive public image for the firm, especially within the aviation community. The brothers had long worried about other aviators gaining commercial advantages by misappropriating their work. Their decision to take the long, tiring railroad trip through the Appalachians to Old Point Comfort, at the tip of the Virginia Peninsula, and to then pick up a steamer to reach Kill Devil Hills was made with secrecy in mind. While the windier shores of Lake Michigan, where Octave Chanute tested gliders in 1896, were closer to Dayton, they were also closer to the major media center of Chicago. The Outer Banks were further from potentially prying eyes, and after writing to the U.S. Weather Bureau, the brothers decided to test their gliders and first airplane more than five hundred miles from home. Even after they shifted their proving ground to Huffman Prairie, outside Dayton, in 1904 and 1905, just south of the Simms Station stop on the Dayton, Springfield and Urbana Electric Railway, they built their hangars away from the rail line and attempted to fly as little as possible when the interurban cars passed by, while the U.S. Patent Office processed their application to patent the crucial parts of the 1903 airplane. After the patent office approved their application, in 1906, enforcing its claims became an important aspect of the brothers' lives, existing in a separate sphere from their commercial interests.

In the summer before the company's formation, the brothers demonstrated their airplane before President Taft, the U.S. Army, Postmaster General Frank H. Hitchcock (Katharine noted that she "like[d] him about the best of the cabinet bunch"), and an "enormous crowd" of "at least 10,000" at Fort Myer, Virginia, in fulfillment of their contract with the Signal Corps. Even while the testing proceeded, Katharine, who had traveled east to watch her brothers at work, wrote to their father that the brothers' patent attorney "[Harry] Toulmin was here Sunday on his way to Virginia. The boys are making arrangements to start suit against Curtis [sic] and Herring immediately." Within the month, Wilbur was in New York to do just that, and when the Wright Company acquired the brothers' 1906 patent, it also acquired the legal battles swirling around the document. Katharine noted that her brothers were less concerned with the patent implications of Louis Blériot's flights in Europe, and instead were "glad he beat [Hubert] Latham" to become the first person to fly across the English Channel.[7]

Protecting the value of the Wright patent was a central concern among the new company's executives when they organized the business in 1909, and they could not have been unaware of the effects of George B. Selden's automobile patent (the validity of which was still being debated in the courts) on the automobile industry.[8] The Wright Company believed that their 1906 patent covered the methods of lateral control used by all airplanes and that all other manufacturers owed them royalties on each airplane built. However, the Dayton firm found it difficult to convince other manufacturers of the comprehensive nature of the Wright patent and of their consequential requirement to gain the permission of the Wright Company to build airplanes and pay it a royalty on each airplane produced. Glenn Curtiss, in particular, asserted that the ailerons used on the aircraft his company built were not covered by the Wright patent and refused to pay royalties or a license fee. The Wright Company responded with a lawsuit in U.S. federal court against Curtiss in which an appeals panel agreed in early 1914 that Curtiss did, in fact, infringe the pioneering Wright patent, but the vehemence with which the Wright Company pursued the case damaged its public image. Prominent aviators not connected with the Wrights, including Henri Farman and Louis Blériot, criticized the actions of the brothers and their company, accusing them of overtly pursuing a monopoly on airplane production at the expense of developing better airplanes. Indeed, the creation of a monopoly was something company executives wanted. They asserted that lawsuits against infringers would increase the business reach of the company, and Andrew Freedman believed that the Wright Company would naturally get the customers of any infringers it forced out of business. Lawyer and director De Lancey Nicoll was brazen about the company's initial plans, asserting at its formation to the press, "What's a patent for if not to establish a monopoly?" Nicoll's comment, made in an era best symbolized by the U.S. Justice Department's successful lawsuit against John D. Rockefeller's Standard Oil Company under the Sherman Antitrust Act for its monopolistic practices—a lawsuit that resulted in the oil giant's breakup into multiple smaller companies—was remarkably blind to the negative image of monopolies held by much of the public.[9]

The Wrights themselves initially held a more diplomatic position than the former New York district attorney, stating that they merely wanted recognition and appropriate compensation for their invention and did not "desire to restrict the use of aeroplanes," and biographer Tom Crouch has suggested as much, noting that the Wrights "wanted nothing more than to be recognized as the true inventors of the airplane," while Glenn Curtiss "wanted

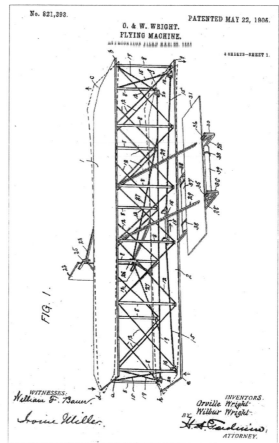

FIGURE 4.2. The first page of the Wrights' 1906 patent application for their flying machine. *U.S. Patent and Trademark Office via Google Patent Search*

nothing more than to develop, build and sell improved aircraft." Curtiss himself once told the Wrights as much, sending the brothers a postcard from France in 1909, where he was competing for the prestigious Gordon Bennett Trophy, in which he wished them "success in every thing except your suit against me." However, the refusal of Curtiss and other aviators to negotiate to the Wrights' satisfaction or to provide the wanted and required recognition and compensation hardened the views of the Wrights, who gave their full support and much of their time to the patent enforcement lawsuits.[10]

President Wilbur Wright assumed principal responsibility for working with lawyers and attending court sessions on behalf of the company. Such work was the president's personal business; in 1910 and 1911 he especially did not want general manager Frank Russell communicating with company counsel. By the summer of 1911 Wright, justifiably frustrated with what he saw as blatant infringements of the 1906 patent by Curtiss and by touring European aviators Louis Paulhan and Claude Grahame-White, and unhappy with the company's balance sheet, looked to patent infringement lawsuits as the best way to save the business—and for the Wrights to relieve themselves of their corporate duties. He wanted the company to exit the exhibition business ("only big profits and a quick release from worry could compensate for putting up with it at all") and "reduce expenses to the lowest possible point" to obtain "a real profit on every machine." Financially sound, the company would then wait for the courts to close its competitors and "do a modest business sufficient to pay reasonable interest on the actual investment and our responsibility will be ended." The Wrights could then leave the courtroom and the boardroom and return to where they truly felt comfortable, the workbench.[11]

Wilbur was not unfamiliar with bitter controversy and intrigue. As a young printer, he had supported his father as he led a schism within the Church of the United Brethren in Christ. The church and Milton Wright had long maintained a vigorous opposition toward secret societies such as the Masons, believing that such groups were unchristian. But by the 1880s, opinions in the church were changing, becoming more accepting of secret groups. Wright's opinions, though, did not change, and church leaders of the liberal faction deposed him from his role as editor of the denomination's newspaper, *The Religious Telescope*, in Dayton and made him the bishop responsible for California, Oregon, and Washington State. The Liberals also proposed a new church constitution that would be submitted for approval at the denomination's 1889 general conference. Milton Wright, though spending much of his time supervising congregations in the West, and worried over the health of his wife, Susan, who was dying of tuberculosis, worked with other conservative church leaders (known, confusingly, as Radicals) to fight the new constitution, and led the creation of a new, conservative United Brethren newspaper, *The Christian Conservator*. Wilbur, at home in Dayton, became his father's adjutant. He wrote and published a short tract supporting the Radical position, based on notes he took while attending a church commission meeting in Dayton while his father was in the West. Through the *Conservator*, the twenty-one-year-old Wilbur's writing reached the multitudes at local church conferences and individual congregations. He also composed editorials for the *Conservator*, attacking those who attacked his father. But his efforts were in vain. At the 1889 general conference in York, Pennsylvania, the delegates overwhelmingly approved the new church constitution and creed by a vote of 111 to 20. Milton Wright was the only bishop voting against the new constitution. Rather than accept defeat and submit to the new order, Wright and fourteen of the twenty "nays" left the conference and moved to a new location in York, where they proclaimed themselves the true Church of the United Brethren in Christ (Old Constitution), to differentiate themselves from the larger group. Milton returned to Dayton as the leader of the Old Constitution church. Over the next several years, he would use the courts in an attempt to maintain control of the United Brethren publishing agency, which was headquartered in Dayton, but the courts sided with the New Constitution church, handing down a final verdict against Milton and his supporters in 1895. Milton remained a leading figure in the Old Constitution church until his retirement, in 1905, loyal to the 1841 constitution and to the plain fashion he believed it represented, even as his children groomed and dressed themselves according to more popular trends. The larger church after the schism, the Church of the United Brethren in Christ (New Constitution), merged with the Evangelical Church in 1946 to become the Evangelical United Brethren Church, which in turn joined with the Methodist Church in 1968 to form the United Methodist Church. Milton Wright's church remains today the Church of the United Brethren in Christ, no longer needing to refer to its constitution in its name.[12]

Recalling his father's years spent fighting unsuccessfully in court for his side of the United Brethren church, Wilbur Wright wanted to pursue infringers as speedily as possible. He found the pace of court proceedings unfriendly to his business's success. Writing in a frustrated tone to patent lawyer Frederick Perry Fish in May 1912, shortly before he fell ill with the typhoid fever that killed him, Wilbur complained that "unnecessary delays by stipulation of counsel have already destroyed fully three-fourths of the value of our patent. The opportunities of the last two years will never return again." The company sued parties large and small, from Glenn Curtiss's companies to individual exhibition pilots such as Claude Grahame-White of Britain and Louis Paulhan of France, and spared no expense. But the suits proved expensive in the court of public opinion. Legal scholar Herbert Johnson suggests that the lawsuits hurt the company financially by alienating wealthy members of the Aero Club of America such as Elbert Gary of U.S. Steel, brewer Gustav Pabst, or businessman and real estate mogul John Jacob Astor IV, who might otherwise have joined Vanderbilt, Belmont, and Freedman by investing in the Wright Company. According to Johnson, "private investor and consumer confidence was shaken by the pendency of the aviation patent suits, depleting non-governmental encouragement for aeronautical development." The cost of patent litigation—at least $28,000 over five years, according to the company's ledger, though the Dayton press placed the figure closer to $100,000, and a Curtiss biographer claimed $150,000—diverted to its lawyers funds that the company could have used to further refine or market its products. The Wright family also believed that stress from the patent litigation weakened Wilbur, making him more susceptible to typhoid fever.[13]

In death, Wilbur Wright received favorable coverage. No obituary writer was going to besmear him over the patent battle. And in Dayton, no journalist was going to besmear him at all. The Wrights, hometown heroes, received favorable coverage in the local press throughout their court battles, especially from the *Dayton Daily News*. The *Daily News*, published by Ohio governor James Cox, who would be the Democratic nominee for U.S. president in 1920, approved of a favorable decision by courts in France in 1911, noting that "the companies that have been [sued] by the Wright Brothers have contended that the Wrights were not the pioneers in aviation, but that they had been preceded by inventors and aviators in France. The French courts showed conclusively that the Wrights flew in 1903 and that no Frenchman flew before 1905 or 1906." When the judges of the U.S. court of appeals issued their January 1914 decision affirming Judge John Hazel's ruling that the Wrights were indeed the inventors of the wing-warping method of lateral control and were the first to fly, the *Daily News* noted with pleasure that the case was the "most important suit yet decided" and that the "Dayton men" were finally legally acknowledged as the designers of the airplane. Dayton's other principal newspapers, the *Dayton Journal* and the *Dayton Herald*, while broadly favorable in their coverage of the Wrights, provided less coverage of the patent battles than did the *Daily News*.[14]

Press coverage of the patent lawsuits away from southwestern Ohio was not as favorable to the brothers or the company. To the *Boston Evening Transcript*'s aviation columnist (likely Franklin Jordan, its aviation editor), the lawsuits made the Wright Company—and Orville Wright in particular—appear to be "determined to kill" the aviation industry and Curtiss, who supposedly ran "one of the most progressive and wide-awake concerns in the country." Its demands for royalties of 20 percent of an airplane's purchase price were "away out of the question." The issue of royalties also concerned a "prominent motor maker," who criticized the Wright Company's conservatism. Writing anonymously in *Aeronautics*, he noted that while he supported a "moderate royalty," the Wright Company was not "progressive enough to be satisfied with a moderate royalty." Meanwhile, aviator and airplane builder Thomas W. Benoist appeared troubled by the 1913 district court ruling in favor of the Wright Company that the U.S. Circuit Court of Appeals for the Second Circuit upheld the next year. While asserting that he did not "think that the decision of the Wright case, one way or the other, will have any particular effect on the American aeroplane industry," Benoist also believed that "if the aeroplane industry in America were restricted . . . to one concern and one head, it would never get very far; while foreign companies, not handicapped by a monopoly, would soon leave this country so far behind that it could never catch up." Alan Hawley, president of the Aero Club of America, urged that aviators accept the verdict of the court. The judge, he wrote, "was unbiased and gave fair consideration to the case before him." An equivocal New York press addressed the decision several times in 1914. The *New York Times* editorialized that while the court order would "restrict the number of experimenters in aerial navigation," the public had "no excuse for assuming that Mr. [Orville] Wright will abuse any power which the law gives him or that he will impose prohibitive charges and conditions on all aviators except his own." Two months after the court issued its verdict, the *New York Sun* wrote that "aviation has been dormant in this country" as a result and that "everyone connected with the industry has been fraid [*sic*] to move." At the end of the year, the *New York Tribune* claimed that the industry's dormancy continued and that the "Wright-Curtiss fight has prevented manufacturers from extending their business and retarded development" throughout the United States. Forming the monopoly that some *Aeronautics* readers feared was very much on the minds of the Wright Company's directors, but not on Orville Wright's. More cognizant of what John D. Rockefeller had undergone with Standard Oil, Orville disagreed with their strategy and bought out the stock of the remaining directors in the months after the release of the appellate decision.[15]

Did the patent battles between the Wright Company, Curtiss, and others play a significant role in the technological stasis that afflicted U.S. aviation in the prewar years? The journalists writing in the *Sun* and the *Tribune* certainly thought so, and many scholars agree. Herbert Johnson, as noted earlier, believes that the lawsuits retarded growth and development in the industry, though the U.S. government's limited financial support of the field also retarded innovation. Industrial sociologist Thomas McFadden believes that private hoarding of patents slowed production, kept potential builders from entering the industry, stifled innovation, and kept the industry as a whole "from meeting market demand." Curtiss scholar John Olszowka notes that the reluctance of the U.S. military and Congress to become significantly involved with aviation, or to perceive aviation as something of value, obstructed aeronautical growth, while also agreeing that the patent lawsuits "stunted the technological development" of the field. Wright biographer Tom Crouch writes that not only did the lawsuits prevent the Wright Company from becoming a profitable industry leader but also the company itself suffered as the Wrights "paid far more attention to winning victory in the patent suit than they did to the development of new and improved product." Still, Crouch doubts that the battle had much lasting damage on the industry. The profitability of the Curtiss Company "and the essential failure of the firm which pursued the case [the Wright Company], is surely proof that the suit was not a significant factor retarding the growth of American aeronautics," he asserts. Instead, Crouch ascribes the anemic state of the U.S. aviation industry to an atmosphere completely different than that which existed in Europe, where "strenuous competition between a relatively large number of designers and aviators . . . led to the exploration of a wide range of configurations, the use of new materials, and improved control systems and power plants." Governments in Europe, where war was brewing, also funded aeronautics much more robustly than did their counterparts in Washington.[16]

There is no doubt that the quantity and value of government marks, francs, pounds, and rubles far outpaced the dollars allocated by the U.S. Congress and significantly helped develop the European industry more quickly than its American cousin. However, the patent war itself also retarded technological innovation, especially any that might have come from the Wright Company. Orville Wright himself admitted as much, telling reporters that the company had "held in abeyance" "radical" improvements in airplanes while the lawsuits progressed. Wright feared that substantial changes to his company's airplanes would be admissions of inadequate design, and he did not want to weaken the claims his lawyers made in court. And court was expensive. The Wright and Curtiss companies, in particular, spent large amounts of money for attorneys' fees, dollars they could not spend on research or development, marketing, or expanded production. By lagging in making or incorporating significant innovations, U.S. companies gave the government little reason to view aviation as anything more than a curiosity or to order significant numbers of aircraft. While the Wright Company had every right to enforce a valid patent, the lawsuits did little to promote aviation's usefulness to those who could have provided significant financial support. Frustrated with the inertia of the company and its place in the industry in early 1914, Theodore Shonts, Andrew Freedman, and other Wright Company investors were happy to sell their stock to Orville Wright. By the end of the patent battle, rich financiers no longer viewed the Wright Company as an investment vehicle. And while Curtiss found much greater commercial success, the fight proved financially costly to his company (which spent nearly $175,000 protecting itself, against the $28,000 to $150,000 spent by the Wright Company), and made it undertake activities of dubious ethical propriety such as the "restoration" and flying of Samuel Langley's 1903 Aerodrome at Keuka Lake in 1914 in its unsuccessful attempt to claim that the Wrights were not first to invent an airplane capable of flight, invalidating their 1906 patent and any possibilities of infringement. Curtiss would be fortunate to stave off enforcement of the court decision through a combination of a lawsuit over a different part of the patent, Orville Wright's increasing disenchantment with corporate life after 1914, and U.S. entry into the First World War, which spurred the creation of an airplane patent pool by the federal government.[17]

Barnes, Freedman, and other Wright Company officers hoped that the increased number of orders, royalties, and license fees it stood to collect through an adjudicated, enforced patent would compensate for the significant amounts of money and time spent on lawyers and in courtrooms in Ohio and New York. Had the Wright Company maintained a 20 percent-of-purchase-price royalty on the hundred or so airplanes it sold to the U.S. military and to private individuals (most of which it priced between $5,000 and $7,500), it would have added more than $100,000 to the company's coffers. Determining the amount of money the company could have raised through license fees is more difficult, since the Wright Company did not license any outside producers to copy its airplanes between its ending its relationship with the Burgess Company, in Marblehead, and Orville Wright's sale of the Wright Company, in October 1915. Also, any fees accruing to the Wright Company from a license agreement would have been individually negotiated. The company's ledger suggests that it collected only approximately $50,000 in royalties during its time in Dayton, making the win before the U.S. court of appeals a pyrrhic victory in light of its effect on subsequent revenues. Its short-lived exhibition department was nearly four times as profitable and proved much more entertaining to the general public than an appellate brief.[18]

Col. Rodgers 1911

# 5

# World Records for Wright Aviators

## *The Exhibition Department*

The Wright Company did not want its only appearances in the press and in the public imagination to be connected with its lawsuits. It hoped to attract positive press coverage (and drive sales) with its exhibition department, and so it hired young, daring, if not foolhardy men as company pilots to demonstrate the capabilities of Wright Company airplanes before the public at fairs and other events. It used the department's performances as advertising, both for aviation in general and for the Wright Company specifically. In 1910 and 1911 the exhibition department was also the largest single consumer of

Wright Company airplanes. The exhibition aviators were the traveling public face of the company (the Wright brothers themselves rarely accompanied them to their performances), one greatly valued by company directors. Alpheus Barnes wrote Andrew Freedman in early 1911 in support of bonuses for the pilots: "The reputation they made for the Wright machine during the past year I consider a valuable asset which more than offsets the amount appropriated." The department grew from a realization by the Wrights—who disliked the showy and dangerous flying common to the aviators—that, in addition to sales, fees for exhibition performances were one of the best sources of income for the company.[1]

To the editors of the *Cameron County Press*, in rural Emporium, Pennsylvania, exhibition aviators foretold the aeronautical future: "bringing flight into the place of first importance." And there were many daredevils flying those early airplanes. The Wright Company's pilots were part of a crowded community of aviators who performed for the public. Capitalizing on the novelty of the new technology, pilots—mostly young, white men, though several women gained fame as pilots—flew airplanes built by all the major manufacturers of the day at provincial, state, and county fairs and public commemorations and competed for money and trophies endowed by prominent aviation financiers such as Wright Company investor Robert Collier (whose Aero Club of America Trophy, first awarded in 1911 to Glenn Curtiss and renamed the Collier Trophy in 1922, remains a prize for which modern aviators compete). While some aviators flew independently, unconnected with a company, during the first half of the 1910s most exhibition pilots were employees of an airplane builder, demonstrating their company's airplanes to audiences from Manitoba to Miami. The Curtiss Company employed such prominent aviators as Lincoln Beachey, Charles Hamilton, and Blanche Stuart Scott, the first woman to fly an airplane by herself; Curtiss himself also occasionally flew. Siblings Alfred, John, and Matilde Moisant flew both Blériot and Moisant airplanes; their stable of aviators included Harriet Quimby, the first woman licensed as a pilot in the United States. While exhibition aviation garnered its sponsors significant financial rewards, it proved quite dangerous to its practitioners, with pilot (and sometimes passenger) deaths from crashes or from losing consciousness at high altitude and falling from out-of-control airplanes without seat belts chillingly common.[2]

The Wrights did not want to personally manage exhibition aviators, and they found it easier to hold to this goal than their initial goal of not personally interfering in the overall management of the company. Even before the company's formation, they received offers from other businessmen, including engineer, Toledo Mud Hens owner, and airship promoter Charles J. Strobel of Toledo, to arrange and supervise the crowd-pleasing shows for which the exhibition pilots became famous. During the summer of 1909, the brothers wrote to one of Strobel's former aviators, the well-known airship builder and pilot Augustus Roy Knabenshue (1875–1960): "We are not thinking of connecting ourselves directly with the exhibition end of the business, but in return for an order for a number of machines we would agree to assign you . . . exclusive rights to give exhibitions . . . during the years 1909 and 1910." Originally from Lancaster, in central Ohio, Knabenshue was the son of Salome and Samuel Knabenshue. His father, a former school superintendent and editor of the *Toledo Blade*, had joined the U.S. diplomatic corps in 1905 as consul in Belfast, Ireland; while their son was gaining fame in aviation, Salome and Samuel Knabenshue were in the process of moving to a new diplomatic post in Tianjin (Tientsin), China, where Samuel served as consul general until 1914. Another Knabenshue son, Paul, would make a career in the foreign service, sheltering nearly five hundred British refugees in the legation in Baghdad, Iraq, where he was the U.S. minister resident and consul general—the highest-ranking official, in 1941. Roy Knabenshue, however, was drawn to the skies. Working as a manager of the Central Union Telephone Company in Fostoria, Ohio, during the day, the young Roy Knabenshue gained local notoriety for building a cigar-shaped airship in his backyard. He first flew aeronautical showman Thomas Baldwin's *California Arrow* airship at the Louisiana Purchase Exposition in St. Louis in 1904. Returning to Toledo, he partnered with Strobel in 1905 and, with his own *Toledo I*, gained fame flying in Toledo, Detroit, and at the Ohio State Fair in Columbus, initially under the stage name Professor Don Carlos, since his family did not support his dangerous new career. After ending his partnership with Strobel, in 1906, Knabenshue launched his own airship exhibition team, partnering with Lincoln Beachey (who went on to fly for Glenn Curtiss) in 1908. But Knabenshue believed that airplanes, not gas-filled airships, were aviation's future, and he was ready to switch to heavier-than-air flight when the Wright Company came calling, in early 1910.[3]

Roy Knabenshue never ordered any airplanes from the brothers, but in March 1910, as spring began and the company's sales of airplanes remained minimal, Wilbur Wright hired him as general manager of the company's new exhibition department, at an annual salary of $5,000 plus 5 percent of any profits from exhibitions. Given a "free hand" in "the management of [his] end of the business," Knabenshue established the exhibition department's offices on the thirteenth floor of downtown Dayton's United Brethren Building, the tallest building in town, which also housed the headquarters of Milton Wright's former church, two and a half miles east of the factory floor and manager Frank Russell, and a mile east of the Wrights, on West Third Street. One of the lasting legacies of Knabenshue's work for the Wright Company was Mabel Beck. Beck, hired as the exhibition department's secretary in 1910, went on to be Orville Wright's personal secretary from 1912 until his 1948 death. Exhibition appearances proved quite profitable for the Wright Company. In little more than a year of flying (June 1910 to September 1911), the company's pilots appeared at events and competitions from Indianapolis to Montreal, San Francisco to New York, earning a profit of $190,298.28 against total revenue of $272,863.16. But earning that money required plenty of footwork and salesmanship. During the spring of 1911, the Wright Company employed three men "on the road," aside from Knabenshue, who was based in Dayton, to promote and sign contracts for exhibitions, most of which cost their organizers between $2,500 and $5,000. Promoting the Wright aviators attracted colorful characters, including André Roosevelt of St. Louis, a second cousin of former president Theodore Roosevelt. The half-French Roosevelt initially came to the company's attention when he began to take flying lessons on behalf of the Missouri Naval Reserves, which wanted to purchase an airplane—and to pay for it with exhibition flights at small communities throughout the state. Roosevelt did not finish his flying lessons, but he did bring his wife to Dayton during the heat of the summer. Katharine Wright, in her informal role as logistics coordinator for favored company figures, found lodging for his wife, Adelheid, praising her for being a lovely woman (though "not quite so much so" as Frank Russell's wife), and the brothers, helpful if introverted, loaned her $100 in November 1911. André Roosevelt's tenure (and the tenure of the other promoters) with the Wright Company was brief, but he went on to produce movies for Vitagraph Studios; fly over, explore, and photograph Andean volcanoes; and become a leading figure in Haiti's tourist industry before his death, in 1962.[4]

Knabenshue's pilots were a colorful lot. Orville Wright personally trained the first group of company pilots in Montgomery, Alabama, in the spring of 1910, wanting to make sure the company's exhibition arm would be ready to make appearances as the weather warmed in the North. "We must have good men who can get things done without being told every move," he informed his brother from central Alabama. Walter Brookins, a Dayton native who rapidly gained Orville's confidence as "a first class man," quickly learned to fly the Model B, becoming both a show pilot and an instructor at the company's flight school at Huffman Prairie. "Brooky" (as his former Latin teacher, Katharine Wright, would affectionately call him) was not impressed with Frank Russell's work as general manager, and he refused to get out of bed at 11:00 a.m. one March morning in 1911 when Russell called for the twenty-two-year-old at his parents' residence, five and a half blocks north of the Wrights' home. The Wrights provided touches of home to some of their pilots. Returning to Dayton from an exhibition in Salt Lake City in April 1911, at which he collected $10,000 for the company and tonsillitis for himself, Philip Parmelee enjoyed a dinner of noodle soup, asparagus salad, and strawberry shortcake, arranged by Katharine Wright. Hoxsey, a mechanically inclined Illinois native, and Johnstone, a trick bicycle rider on the vaudeville circuit, became the company's Stardust Twins, continually trying to best each other in feats of altitude and acrobatics. Their antics caused the wary "Wrights repeatedly to remonstrate with both on the desperate chances they were taking." In fact, many early exhibition pilots did not live to see the start of the First World War. They did not even live long enough to attend Wilbur Wright's funeral. Parmelee died in a crash at an exhibition in Yakima, Washington, just two days after Wilbur's death, in Dayton in 1912. Large crowds, Frank Russell noted, made "none but the most

FIGURE 5.1. Exhibition department pilots Arthur Welsh and George Beatty in a Model B, 1911. *Courtesy of the Library of Congress*

conservative" of pilots "able to resist attempts at something more sensational than rivals have to their credit." Hoxsey and Johnstone died in separate accidents while performing in 1910. Parmelee and Hoxsey were twenty-five when they died, and Johnstone twenty-nine. Brookins fared better than several of his colleagues, dying at home after a long illness in 1953.[5]

The pilots' death-defying twists, turns, and corkscrews in open fabric-and-wood airplanes brought paying customers to fairs and festivals—for a while. The income from the Wright Company's exhibition department—nearly $4.5 million in 2010 dollars—greatly helped the company's balance sheet. Nevertheless, over the 1911 exhibition season, Glenn Curtiss's exhibition company—which sent its pilots to 210 cities (the Wright Company's pilots appeared in only 38 cities that year) and accounted for nearly 80 percent of exhibition flying in North America—provided the Wright Company stiff competition for exhibition revenue. With the initial novelty of aviation fading, competition from an increasingly crowded field (which included assorted independent pilots), and eager to avoid the deaths of more of its pilots after Hoxsey and Johnstone died while showing its airplanes, the Wright Company decided to close its exhibition department in the autumn of 1911. Roy Knabenshue, who had spent the winter of 1910–11 in California considering whether he wanted to remain with the Wright Company before returning east for what would be the exhibition aviators' last season, resigned from the company on 1 November 1911, though he remained on friendly terms with Wilbur and Orville Wright. Knabenshue later formed his own company to build airships and, from 1933 to 1944, worked for the National Park Service, developing its plans for using autogyro aircraft in conducting land surveys and spotting and fighting forest fires.[6]

Of course, exhibition aviation did not cease with the end of the Wright Company's exhibition department; throughout the world, air shows remain popular (though sometimes dangerous, even deadly). The exhibition arm of Curtiss, incorporated in September 1910 as the Curtiss Exhibition Company (which became part of the reorganized Curtiss Aeroplane and Motor Company in 1916), continued to operate. But Glenn Curtiss was less cautious than the Wrights. Aviators flying for his Hammondsport company leased their airplanes and were allowed to keep portions of their appearance fees and of any cash prizes they won, while Wright exhibition flyers used company-owned aircraft and earned a flat rate of $20 per week plus $50 for each day flown. Any prize money won by Wright aviators went to the company's treasury, and, in keeping with the brothers' own personal standards and their upbringing in the house of a leading bishop of the Church of the United Brethren in Christ, they also were forbidden from drinking, gambling, or flying on Sundays. Curtiss aviators faced no such strictures. At any rate, exhibition flying was as deadly for Curtiss aviators as it was for the Wrights' flyers; Charles Walsh, Julia Clark (the third woman licensed to fly by the Aero Club of America), and Cromwell Dixon all died flying for the Hammondsport firm. "It got so we were losing a pilot a month," Curtiss aviator Beckwith Havens remarked about the industry as a whole. Ironically, Havens would live to see the beginnings of the Apollo program, dying in a Connecticut hospital in May 1969.[7]

Even though the Wright brothers were less than enthusiastic about their company sponsoring exhibition pilots (at one point referring to exhibition aviation derisively as "the mountebank game"), they made certain to offer aircraft specifically targeted to independent pilots. During the company's first years, it offered several models specifically for exhibition aviators who might find a regular Model B—a 1,250-pound craft with a 39-foot wingspan—too cumbersome for their performances. In 1910 the Wright Company introduced its Model R, a smaller version of the Model B; sometimes known as the Roadster or Baby Grand, the Model R had a 26.5-foot wingspan and weighed only 750 pounds. The Wright Company's pilots set several records with the R in 1910. Walter Brookins set three world records for altitude at exhibitions in Indianapolis (the first exhibition by the company's aviators) and Atlantic City and reached 6,175 feet in New Jersey, while Ralph Johnstone flew it to 9,714 feet at Belmont Park on Long Island at the October 1910 International Aviation Tournament. Reporting on events at Belmont Park, Katharine Wright, in New York with her brother, told the bishop that Orville was speeding through the coastal sky. "Orv took our big (or little) racer and made almost seventy miles an hour," she wrote, a speed over two times as fast as the maximum speed of 30 miles per hour for the 1903 airplane. Speed, not safety, would be the marketing focus for the Roadster. The Wright Company advertised the R as "the fastest climbing aeroplane ever built," offering it at $5,000 in 1911. It is unlikely, though that the company found much success in convincing aviators outside its employ to acquire the aircraft. Company ledgers provide no data on specific sales, and contemporary newspaper articles about exhibitions often note only the make of airplane flown by an exhibition pilot (though some exhibition aviators built their own airplanes, in the style of one of the larger companies); they are less likely to specify the exact model a pilot flew. With the Wright Company's limited overall production, few pilots not managed by Roy Knabenshue awed crowds by flying the Model R.[8]

The Wright Company also tried to market the Model EX to exhibition pilots in 1911, gaining some attention with one notable pilot. Taking its name from *exhibition*, the one-seat EX had a wingspan of thirty-two feet, slightly larger than the R (which was built for racing more than for extended flying), and was the first airplane manufactured exclusively at the new factory off West Third Street. The EX was more powerful than the standard Model B; Wilbur Wright noted in a letter to army aviator Charles Chandler, "Our estimates show that the 'EX' machine with six-cylinder motor would have a speed of nearly 65 miles an hour, with one man, and would climb nearly 600 feet a minute." The EX is best remembered for the first flight across the United States, when Calbraith Rodgers flew the *Vin Fiz* from New York to California between September and November 1911, making more than seventy stops. Philip Parmelee also flew an EX at a Saint Louis air show in October 1911. While the company hoped these flights would demonstrate the airplane's positive attributes and create orders for the Dayton factory to fill, the exploits of pilots in the EX generated little other attention and no obvious orders from outside aviators, even after Rodgers's journey from the Atlantic to the Pacific. Rodgers's difficulties in making the transcontinental journey did not enhance the Wright Company's reputation.[9]

The cigar-chomping, teetotalist Rodgers (1879–1912), originally from Pittsburgh, was related to famed U.S. Navy commodores Oliver Hazard Perry and Matthew C. Perry and dreamed of naval service. In 1911, his cousin John Rodgers, a navy lieutenant, learned to fly at the Wright Company's school at Huffman Prairie. Cal Rodgers, who could not follow his family's tradition of naval service due to profound deafness from a childhood bout with scarlet fever, visited John Rodgers in Dayton, where he developed his own interest in flight and enrolled in the Wright Company school, soloing in a Wright B for the first time that August. Confident of his skills as an aviator (and not as affected by the loud drone of the engine as others) and ineligible to fly for the military, Rodgers decided to pursue the potential of fame and fortune offered by *New York Journal* publisher William Randolph Hearst. Hearst, famed for bitter battles with *New York World* publisher Joseph Pulitzer that led to the epithet *yellow journalism*, believed that aviation needed fewer exhibitions at state fairs and more and better airplanes. He hoped that his prize would encourage the construction of stronger, faster airplanes—and that readers would buy Hearst newspapers to read their coverage of the competition.[10]

Rodgers's flight—one of three recognized attempts to win the $50,000 Hearst Transcontinental Prize for the first person to complete a flight across the United States within a thirty-day period during 1910–11—served as an informal advertisement for the Wright Company, since thousands of people saw his airplane at his various stops, but it contractually served as advertising for meatpacker Armour and Company of Chicago. Armour, which had recently introduced Vin Fiz, a grape-flavored soda, hoped that Rodgers's airplane would serve as a flying billboard; Rodgers, who cared little for the soda (its syrup tended to spoil in the unrefrigerated rail cars that transported it), was expected to refresh himself with the drink after landing, though he preferred to drink cream. Rodgers's support crew—which included longtime Wright brothers and Wright Company mechanic Charles Taylor, officially on a leave of absence from the Dayton firm—painted "Vin Fiz" on the underside of the airplane's wings, and Rodgers himself dropped advertising leaflets from the air as he overflew gathered crowds. Flying as an independent aviator, Rodgers purchased his airplane for $5,000 and also bought $4,000 of spare parts that accompanied the flight on a support train. Leaving New York on 17 September, Rodgers reached Pasadena on 5 November. When he reached California, nearly three weeks after the thirty days specified by the Hearst prize, only the vertical rudder and drip pan remained of the airplane that left New York.

FIGURE 5.2. Cal Rodgers (*second from right*) showing off the *Vin Fiz*, 1911. The others in the photograph are unidentified. *Courtesy of the Library of Congress*

Rodgers found the trip across North America costly. He endured sixteen separate crashes (some of which gave him serious injuries), and *World's Work* writer and editor E. French Strother estimated that he spent between $17,000 and $18,000 in airplane repairs alone. Other daily necessities added to the total. Armour did pay Rodgers—$5 per mile flown east of the Mississippi and $4 west of the river—but the pilot also incurred at least $1,300 in royalty charges to the Wright Company while on his trip. Wilbur Wright was sufficiently concerned about Rodgers's ability to pay his bills that he directed the factory to not ship the pilot "large orders of parts" without first receiving payment. While Rodgers demonstrated great personal perseverance in completing the trip after he could no longer win the $50,000 prize (he had only reached Kansas City when the thirty days expired), his expedition also demonstrated the fragility and limited endurance of period airplanes. The stronger machines Hearst hoped for were still a dream. The *Vin Fiz* flight succeeded as a spectacle, but it brought little new business through the doors of the Dayton factory.[11]

While the Wright Company continued to advertise in the aviation press that its airplanes were suitable for exhibition use, Rodgers's flight was the last exhibition of a Wright airplane that garnered attention in major metropolitan newspapers. The company itself finally decided that the costs of supporting its own exhibition flying squad outweighed the potential (but unrealized) public-relations benefits. The aversion of the Wrights to exhibition antics and a general decline in the novelty and profitability of exhibition flying contributed to this decision, as did pilot mortality. Over the seventeen months of the exhibition department's existence, two of its pilots—Johnstone and Hoxsey—died in crashes, and three others who flew for it died in airplane accidents before the end of 1912, leaving just four of the original nine men alive. The company carried no insurance on its aviators, but the executive board voluntarily established annuities for Johnstone's widow and family and for Hoxsey's mother. Having entered the exhibition business with limited enthusiasm, the Wrights had finally lost interest altogether, as had most of their exhibition flyers. At the *New York Times* Aerial Derby in October 1913, only one of the seventeen entrants flew a Wright model. Instead, the airplanes flown ranged from factory-built Curtiss biplanes to the home-built tractor biplane of O. E. Williams of Scranton, Pennsylvania. Five entrants flew French models, and no one seemed concerned about potential patent infringement. The anticipated Wright monopoly of 1909 was nowhere to be seen.[12]

With the end of Wright Company–sponsored exhibition flying, in November 1911, the company's best revenue stream dried up. The company's one attempt at vertical integration also collapsed; it no longer could sell itself its own airplanes. And the company had been its own best customer. The department's pilots had flown a total of twenty-one airplanes, destroying eleven in accidents, planes that cost the company $1,800 apiece to produce (suggesting that the Wright Company intended to make a profit of at least $3,200 on the $5,000 list price of Model B airplanes that were sold to the public and to the government). Given the novelty of aviation, the Wright aviators attracted extensive media coverage and rapt crowds when flying, bringing "looks of wonder on the faces of the multitude. From the grey-haired man to the child, everyone seemed to feel that it was a new day in their lives." One reporter described an airplane's takeoff as "the moment of miracle." However, performance before crowds across North America did not translate into additional sales of Wright airplanes, or help make aviation an enterprise with practical civil applications in North America. By August 1912, with the exhibition department closed, sales were so slow that Alpheus Barnes wrote to one of De Lancey Nicoll's law partners, John D. Lindsay (no relation to the future New York City mayor), "The prospects of a dividend look pretty slim for this year. The long series of fatal accidents since the middle of May [especially of aviators flying the Model C] has caused a depression in interest which will hardly pass away in time to do much business this season." While the Wright Company began its existence with a competitive range of products, by the time Barnes wrote his letter it was finding selling its different models of aircraft to be a difficult proposition.[13]

# To Change or Not to Change

*Creating New Airplanes and New Pilots*

Not only did the Wright Company lose a valuable revenue stream when its exhibition department closed, but also it lost a convenient way for its pilots to test new technologies and designs in the rough and tumble of field use before incorporating them into production models built for public sale. This loss gained little notice in company communications, though, and at any rate the company developed few novel aeronautic technologies or airframe designs in need of the sorts of testing that the exhibition pilots might have provided. While the prototypical airplane changed significantly between 1910 and 1915—from pusher biplanes, where the pilot sat on the lower wing, to tractor biplanes and monoplanes with partially enclosed cockpits—the prototypical Wright Company airplane remained much the same, with the most significant changes in the company's models (even its entrants in the seaplane category) concerning airplane size and engine power, not fundamental design. The company's reluctance to keep pace with the innovations of other producers and to incorporate innovations of its own into its aircraft also reduced its sales—neither the military nor private aviators wanted to pay new-airplane prices for yesterday's technology.

Indeed, the Wright Company pointedly resisted making significant modifications to its airplanes while the adjudication of the 1906 patent wound its way through the courts. The instruction to limit the incorporation of new technology came from the company's highest level: its president. Orville Wright admitted in 1914 that the patent case contributed to the company's inertia, telling a *New York Herald* reporter that, with the scope of the patent seemingly settled, the company was "ready to go ahead with improvements in the aeroplane now. Many of these, some radical in character, have been held in abeyance while our rights were in question. It was important to obtain protection first before bringing them before the world." Wright was unwilling to acknowledge the many technical improvements implemented in airplanes throughout the world in the previous years, including single tractor propellers instead of dual pushers, and intuitive control yokes instead of the Wrights' harder-to-learn lever system, asserting that the airplane of 1914 was "dynamically . . . just the same as it was ten years ago. This is as true abroad as it is in America." His company's reputation as being behind the times technologically even reached Europe. In early 1914, Grover Loening, then working in Dayton as the company's manager, wrote to French engineer, aviation writer, and onetime Farman Company of America president Ladislas d'Orcy to counter that perception, advising him "that the criticism that the Wright Company does nothing to advance the science is now most unjust, since practically 90% of the time and labor of everyone connected with the Company at present is in the most advanced development work that I believe has ever been undertaken by any aeroplane company in the last four years." Wanting to portray his employer in the best light, Loening bragged that the Wright Company was building new airplanes for the U.S. Army and Navy, though the army accepted (and subsequently dropped from its inventory) only one Wright airplane and the navy none after his letter arrived in France in 1914. By 1914, whatever attempts the Wright Company made to improve their airplanes were ignored by their potential customers.[1]

While Orville Wright wanted the company to move forward with the development of new, commercially competitive models after its victory in the courts, finding the space for innovation to occur proved difficult. The company's factory provided no space dedicated to research and development, and 1914's limited income did not allow for anyone to even consider adding another building to the complex. Instead, the company's experimental drafting and design occurred at the brothers' former Wright Cycle Company building, at 1127 West Third Street, nearly a mile and a half to the east. Though not formally company property—the Wrights personally leased it from Charles Webbert until 1916—the structure had been informally used for company business from the start. The Wrights converted the second floor of that building into office space in 1910, and they had maintained their personal offices there since the company's formation. During the firm's first year, company employees built engines in the former bicycle showroom on the first floor, since the rented space at the Speedwell plant was too small to accommodate both motor construction and airframe assembly. After the west Dayton factory opened, the first floor at 1127 West Third became the company's experimental office space, while company pilots tested experimental airplane designs at Huffman Prairie (or on the Great Miami River with the Model G Aeroboat and the Model CH floatplane) before starting general production. But the space was more designated than actual. Notebooks and records of experimental staff do not survive, and the letters that do exist ignore the topic.[2]

Moreover, there was no staff dedicated to research. The Wrights were unwilling to let others modify their basic biplane design, and no patentable ideas proceeded from the drafting table, in west Dayton, to Huffman Prairie or the factory. No one in the Wright Company's employ kept patent examiners' lights burning late into the night. The one idea sent on to Washington for consideration, which Orville tested in his famous 1911 glides at Kitty Hawk, was an automatic stabilizer for which the brothers filed a patent in 1908, predating the Wright Company's formation. Indeed, none of the U.S. patents granted to the Wright brothers during the Wright Company years covered inventions developed on behalf of the firm, nor was Orville Wright's single post-1915 aviation-related patent—issued in 1924 for a split wing flap on behalf of the Dayton-Wright Airplane Company—filed during the Wright Company's years in Dayton.[3] Nor did other Wright Company employees receive patents for inventions developed on behalf of the company. The Wrights may have claimed that they wanted to retire to experimental work, but actually conducting that work to the point of filing patent applications was never a priority for either Wilbur or Orville between 1909 and 1915. No new patents granted also meant that there were no new patents that the company could license to other firms for additional revenue. During the company's first years, Orville Wright principally worked at training new aviators at Huffman Prairie, on testing his stabilizer, and, after Wilbur Wright's death, on the lawsuits, which had enveloped the elder brother until his death. The company's ledger contains several pages of entries concerning patent litigation but contains no pages specifically related to experimentation or research and development, even after the U.S. military all but ceased purchasing Wright airplanes in 1912 after accidents with the Wright C killed several army aviators. Nor did the Wright Company undertake significant efforts to incorporate into its planes technical advances made by other manufacturers, which could have demonstrated weakness in the courtroom.[4]

Meanwhile, patent examiners spent more time between 1909 and 1915 reviewing applications filed by Glenn Curtiss—who was certainly aware of the legal and financial value of an adjudicated patent—and other Curtiss Motor Company employees. Curtiss himself was listed as inventor or coinventor on more than thirty aviation-related patents filed during those years, though he assigned most of his inventions to his company as a whole. These patents covered various hydroplane designs, airplane control mechanisms, landing gears, and other devices. Whether or not Curtiss had a personal role as an inventor or merely was listed on the patents in his role as his company's president, these filings show that his corporation maintained an active research-and-development program. Nor was Glenn Curtiss the only person connected with the Curtiss Motor Company to obtain a patent. Engineer Henry Kleckler, a close Curtiss confidant, and John P. Tarbox, a patent attorney who later served as a director of the Manufacturer's Aircraft Association, also received patents for innovations developed for the firm. Curtiss's pursuit of large numbers of patents was unusual among other prominent U.S. airplane builders. While the patent office awarded plenty of inventors with rights for a variety of developments related to aeronautics in the roughly thirty-eight thousand patents it awarded each year between 1910 and 1915, few of those inventions came from individuals connected with the era's prominent companies. Even Starling Burgess's firm, in Marblehead, was invisible to the patent office in Washington, reappearing in its files only after its purchase by Curtiss, in 1916.[5]

The Wright Company was unable to incorporate new, patentworthy innovations into its products, and sales of its increasingly obsolescent airplanes fell markedly after 1911. The company's ledger shows a marked decline in deposits made on purchases of aircraft after 1912 (and a decline in the level of detail given on the sales and accounts receivable pages). But sales had not always lagged. In addition to the airplanes used by the exhibition pilots, at least fifteen

Model B airplanes had left the factory as private purchases in 1911, as did Cal Rodgers's *Vin Fiz* EX, when the model was still a competitive piece of technology. Using a 10 percent executive discount, a few company officers, including Robert Collier and Russell Alger, purchased Model Bs. Collier, though, never personally flew his new airplane, but it became moderately famous nonetheless. He quickly put it to work by loaning it to the U.S. Army Signal Corps for Lt. Benjamin Foulois to use as an observation platform along the Rio Grande, as he patrolled the border in the wake of the Mexican Revolution. Collier's airplane, and the anticipated loan of five or six others (a loan, since Congress had not yet appropriated money to buy airplanes) would allow the United States to soar "constantly over the crooked channel of the Rio Grande and back into the country where illegal expeditions may be organizing," empowering "the military to establish an impenetrable patrol along the border." By the fall of 1911, Collier's plane was back in the capable hands of Wright Company pilot Oscar Brindley, who took photographer James Hare into the skies to photograph the Harvard-Princeton football game in New Jersey. When the airplane appeared at "one of the most exciting periods of the game," "everybody forgot the battle below and looked above." Princeton defeated Harvard, 8–6.[6]

Some wealthy individuals also purchased airplanes. Joining the "thin ranks of amateur fliers" between 1910 and 1912 were New York yachtsman, carpet manufacturer, and socialite Alexander Smith Cochran and Fairbanks Scale Company heir Harold Haskell Brown of Brookline, Massachusetts, who acquired their own airplanes from Dayton, as did Grover Cleveland Bergdoll (1893–1966) of Philadelphia. The nineteen-year-old Bergdoll, the "Playboy of the Eastern Seaboard," came from wealth built through his family's brewing company. Eager to learn to fly, the University of Pennsylvania law student wrote the Wright Company in February and March 1912 asking how quickly he could enroll. "Answer at once," he demanded of the company, "so that I can send certified check and . . . cable you when I will be at Dayton." In April he arrived in Dayton to learn to fly from the Wright Company. He returned home a pilot, and with the thirty-ninth Model B built in Dayton. That summer, he flew all around southeastern Pennsylvania from what he called Eagle Field, circling the statue of William Penn atop Philadelphia's city hall and racing a train traveling between Devon and Villanova. He took passengers aloft for fun, and made the first flight between the City of Brotherly Love and Atlantic City, a distance of nearly 60 miles. But Bergdoll was not just a young man, eager to learn. He also had a darker side, one that put him into conflict with legal authorities on several occasions. In 1913 he spent three months in jail, convicted of a variety of reckless-driving charges. Writing Orville Wright, he mourned that his Model B would be idled and that for aviators, "bad luck seems to be their only lot: If they are not killed they go to jail, if not to jail, they go bankrupt, and so on down the line." Wright wrote a fatherly letter back, noting that he was "sorry to see such a fine bird caged, but sometimes such is necessary." But a few months in the Montgomery County jail, in Norristown, for reckless driving would be the least of Bergdoll's worries. After his release from jail in the summer of 1913, he continued to fly, even attempting to fly a Deperdussin airplane in France in the Gordon Bennett Trophy competition that September. Bergdoll returned to the United States, dropped out of law school, and flew recreationally from his field. When war came to Europe, he jumped at the chance to use his airplane to aid his family's ancestral land of Germany. Bergdoll visited the German consulate to offer his airplane and himself in the service of Kaiser Wilhelm II, but the consul refused his proposal, telling him that a citizen of a neutral country would have to go to Germany to enlist, and that his plane would never be allowed to leave America. In 1914, Bergdoll hangared his plane. He never flew it again. Philadelphia's Franklin Institute acquired it for display in 1935.[7]

In April 1917, almost three years after the start of the First World War, the United States declared war on Bergdoll's beloved Germany and entered the conflict on the side of the Allied Powers. To raise the four hundred thousand soldiers President Woodrow Wilson requested, Congress passed legislation to institute a draft. Bergdoll was drafted. Not wanting to fight Germans, he went into hiding. Caught in 1919, he was convicted of draft evasion and sentenced to five years in prison. But he did not report to prison. Instead, he fled to Germany, where he lived until 1939. As Germany prepared to launch what became a second world war in Europe, Bergdoll returned to the United States. His 1919 conviction still stood, and he had an extra three years added to it for his escape. He served almost five years of his sentence, leaving the federal prison at Leavenworth, Kansas, in February 1944. Orville Wright refused to omit Bergdoll's name from the list of students taught by the Wright Company to fly at Huffman Prairie that was inscribed on the Wright Memorial on Wright Brothers Hill, outside Dayton in 1939, but he never again met the aviator and draft evader. After his release, Bergdoll moved to Virginia, where he died in 1966.[8]

Bergdoll's airplane, built in January 1912 and purchased that spring, was one of the few private sales made during the Wright Company's final three years in Dayton. The company's ledger suggests that private sales all but ceased after the introduction of the Model C, in 1912. Some sales to the U.S. War Department still occurred, but Wright Company sales to the U.S. Army and Navy were far fewer than those of Curtiss, even amid the patent lawsuit. Between 1909 and 1916 the U.S. military purchased fourteen Wright Company airplanes, only one of which it acquired after 1912. During that same period, Curtiss sold 114 aircraft to the army and navy (a number itself dwarfed by the 5,043 airplanes Curtiss sold to the wartime military in 1917 and 1918). As with the Wright Company, figures for Curtiss sales to private individuals are not extant, nor are comprehensive sales figures for other builders. But Curtiss could survive on its sales to the military. The Wright Company could not.[9]

Collier and Bergdoll purchased their airplanes during the company's best years. An airplane was a useless, large, expensive paperweight, though, if its owner did not know how to fly it. The Wright Company, wanting its airplanes to actually be flown instead of acquiring dust on the ground, found that its flight schools provided another useful revenue stream, one especially needed after the cessation of company-sponsored exhibition flying in 1911. Training programs were common among the prominent airplane companies; they all attempted to capitalize on the airplane's novelty by opening flight schools. In 1911 and 1912 the Curtiss and Burgess companies began to train pilots to fly their planes at their respective flying fields in Hammondsport (and later, for those on the west coast, in San Diego) and in Squantum, Massachusetts, southwest of Marblehead, near Boston. The Wright Company first ventured into pilot education with an informal training camp located near Montgomery, Alabama, at the site of the future Maxwell Air Force Base, from March to May 1910. During those months, Orville Wright worked with five men who wanted fly for the company's exhibition department, providing them instruction in a warmer environment than a Dayton spring. With the advent of warmer weather in Ohio, the company opened the school to the public and moved it to Huffman Prairie to be close to the factory, for easy access to new airplanes and replacement parts. After the exhibition department disbanded, the school provided aeronautical employment for former Wright pilots J. Clifford Turpin and Arthur ("Al") Welsh, who joined Orville Wright (before he became company president) as instructors. Turpin, Welsh, and other instructors at what was variably known as the Wright Flying School, Wright School of Aviation, or Wright Company School of Aviation, trained many famous aviators, including Henry H. ("Hap") Arnold, who commanded the U.S. Army Air Forces during the Second World War; exhibition pilot Marjorie Stinson and her brother, Stinson Aircraft Company founder Edward

FIGURE 6.1. Marjorie Stinson, shown here with a group of U.S. Army officers in 1918, learned to fly with the Wright Company at Huffman Prairie Flying Field. *Courtesy of the Library of Congress*

Stinson; and A. Roy Brown, a Royal Air Force pilot from Ontario credited by the RAF with shooting down "Red Baron" Manfred von Richthofen. The Wrights valued their home life, and occasionally extended it to some of their students who spent weeks away from home learning to fly. During a summer-like May evening in 1911, Arnold, army lieutenant Thomas Milling, and navy lieutenant John Rodgers broke bread with Milton, Katharine, and Orville Wright. Katharine enjoyed the officers' company immensely, telling Wilbur in Germany that Milling, who became the U.S. military's first pilot rated a "Military Aviator," was "a splendid little chap," and Orville enjoyed both his company and Arnold's, even if Arnold wasn't "so attractive" as Milling. Of Rodgers, then still training while trying to cobble together the money for an airplane, Katharine merely noted that he was "the Navy man who has been here for six weeks or more."[10]

As the baking Ohio summer of 1911 turned into autumn and flying in the North again became an uncomfortable activity, the Wright Company closed the Huffman Prairie school for the season and sought to open another school in a warmer climate. It decided on a site near Augusta, Georgia, a site also used that winter for training by the army's Signal Corps. A desire to be closer to a larger center of population led to the opening of a second summer school on Long Island in 1911 at the Aero Club of New York's aviation field; that school later moved to the former site of the Moisant School of Aviation. Orville Wright took special interest in the Dayton school, spending much of his time in 1910 and 1911 at Huffman Prairie flying and helping train many of its early graduates. Only with Wilbur Wright's death, in 1912, did Orville—rather halfheartedly—turn his attention from activities at Huffman Prairie to the Wright Company's business operations. Wright often found traveling long distances difficult due to sciatica resulting from his Fort Myer accident in 1908, and he had little direct involvement with the operations of the Georgia and New York schools. One needed to be wealthy or be funded by the government to afford the steep tuition. The Wright Company charged $500 per student during its first years for a series of lessons, usually over ten days, each lasting between five and thirty minutes; while it reduced the fee to $250 in 1914, when few people bought Wright airplanes, some other schools charged students only $200 to learn to fly. A Wright student would usually have a total of two to four hours of flight time upon completing the course (as did students at other aviation schools). For airplane purchasers, the company offered a special rate of $25 per lesson, resulting in a total tuition of approximately $250. The company's instructors usually taught four or five students at once on the ground, and were well compensated for their work, though, as airline pilots later would be, they were paid only for time spent flying. In 1914 the company hired Howard Rinehart as a teacher at $20 per hour in the air (nearly $450 an hour in 2010 dollars), with an extra $10 per hour if he released the company from any potential liability for accidents or injuries sustained while working. The Wright Company directed him to spend no more than four total hours in the air with each pupil "unless otherwise arranged." If students each paid tuition of $375 (tuition ranged between $250 and $500, depending on the year and whether a student purchased an airplane), the 119 aviators trained at Huffman Prairie earned the company approximately $45,000 between May 1910 and February 1916—a little less than $6,500 per year, and a little more than the cost of one airplane. The schools in Augusta and on Long Island—the only public-facing company operations outside the Dayton area—also added to its bottom line.[11]

Beyond the experience gained while flying with an instructor and flying solo, Wright students at Huffman Prairie practiced using the company's intricate control levers on a grounded Model B at the west Dayton factory. There they also watched workers build airplanes, learned how to reassemble them after storage (since planes were not flown but crated and shipped between distant points), and provided physical labor in positioning aircraft at Huffman Prairie. Students could move airplanes on the ground and in the air without fear of being charged fees for breaking something. While the Wright Company covered the costs of any breakage that occurred during a student's training, not every aviation school was so generous. The enrollment at the company's schools was predominantly male. Though several women—including Rose Dougan, Leda Richberg-Hornsby, and Marjorie Stinson—learned to fly at the school at Huffman Prairie, an advertisement the post-1915 management placed in a trade publication for the Long Island school spoke to a potential student's masculinity, claiming that "flying is the greatest sport of red-blooded, virile manhood" and that learning to fly was an excellent vacation activity. "Live in the open—in the aviator's tent city" near New York, proclaimed the Wright Company. "Convenient hotels for the fastidious," it wrote, suggesting that the grit and grease of the airplanes of the time might make aviation a poor "sport" for those averse to dirt.[12]

Indeed, early aviation was principally a male activity. While women learned to fly at most flight schools, with at least five unmarried women taking lessons at the Martin (Los Angeles), Lillie (San Antonio), and Moisant (Augusta, Georgia) schools during the winter of 1912, they made up a small portion of those learning to fly. Nor did many women work as instructors. Marjorie Stinson, who taught at the flight school she and her sister Katherine established in San Antonio, was one of the few women to teach others how to operate an airplane. A few women also flew in exhibitions, but male exhibition flyers far outnumbered them. But the novelty of a woman being a pilot drew crowds to events where they flew. Blanche Stuart Scott, who exhibited Curtiss airplanes, wrote that her audiences paid "to see me risk my neck, more as a freak—a woman freak pilot—than as a skilled flier." The Wright Company made little effort to welcome women to aviation—as pilots, mechanics, or factory workers—in areas other than the secretarial or needle trades, fields traditionally reserved for women, or to market aviation to women. Only with the advent of the First World War, as men joined enlarged militaries, did women gain employment in airplane factories in trades traditionally reserved for men, and the end of the war reversed those gains as men returned from the European war in search of work.[13]

Whether they were built by men or women, building and selling airplanes— what the Wrights hoped would be the company's principal activity—proved less profitable than exhibition flying and teaching new aviators. Still, between 1910 and 1916, workers at the company's Dayton factory built thirteen different models, eleven of which were built before Orville sold the company. With these airplanes, the company attempted to respond to the needs of prospective customers and to offer a Wright-branded option for the most popular types of aircraft. Between 1910 and 1915 the Wright Company's workers built approximately 120 airplanes in Dayton. This figure included thirteen of the fifty-nine airplanes acquired by the Signal Corps, and two of the twenty-three procured by the navy, during those years. The company's exhibition department used twenty-one airplanes, leaving approximately eighty-four (nearly 70 percent of the company's total production) available for private purchase—though remaining company records do not reveal whether it sold all these models. The Wright Company's total production of airplanes was much smaller than the total production of some of its competitors. While no precise prewar production figures for Curtiss or any other maker exist, the Hammondsport firm's workers produced many more airplanes between 1910 and 1915 than did the Wright Company, building more than 150 aircraft in 1912 alone. Curtiss, more nimble in responding to changes in the market and more willing to change its designs to reflect advances in technology, became the largest domestic airplane builder before the American Expeditionary Forces left for Europe. Resistance to change and to technological innovation in the airplanes that composed the Wright Company stable demonstrate how the company became an also-ran in the U.S. aviation industry.[14]

The various Wright Company airplanes built in Dayton between 1910 and 1915 reflected Orville Wright's technological conservatism. There were few avenues for new ideas to reach the factory floor. The company employed no aeronautical engineers, only draftsmen such as Louis Luneke or Rufus Jones to draw the plans that guided workers during airplane construction. Those plans reflected the ideas of the Wright brothers. In an interview decades later William Conover recalled that "the designing was between Wilbur and Orville absolutely." The company's draftsmen had no independent design role. In replying to a request for an employer reference for Rufus Jones in 1914, Orville Wright noted that while Jones's work was "very satisfactory," he "had little to do with designing." The design that Luneke, Jones, or other company draftsmen copied of the company's first production airplane, the Model B, departed from the style of earlier Wright airplanes, including the 1909 Signal Corps Model A, though it shared the same engine as the airplane flown at Fort Myer that year.[15] It and future Wright Company airplanes no longer used front elevators—the elevator was now at the airplane's rear, behind the pilot. It also landed and taxied on wheels instead of wooden skids and did not require the use of a catapult to get into the air. Historian Richard Hallion considers the B, with its "headless" construction, to be the first "inherently stable Wright machine" since its center of gravity was ahead of its neutral point. The B, principally produced from 1910 to 1912, proved to be the Wright Company's most commercially successful airplane. It was the primary model built and sold during the exhibition department's existence and Frank Russell's tenure as general manager, the model on which students at the Wright schools learned to fly, and the first U.S. airplane manufactured in substantial quantities for purchase by private parties. This open, two-seat biplane with a four-cylinder engine featured "duplicate controls, operable in either seat, an original feature famous in the Wright machines from the beginning." Future Wright airplanes would vary in their number of seats (one or two) and use different engines, but they all followed the basic design of the B.[16]

The B was the airplane used to make the first cargo flight in the United States. In November 1910, company pilot Philip Parmelee flew one from Huffman Prairie to Columbus carrying two hundred pounds of silk for the Home Dry Goods Store for a $5,000 cargo charge. Max Morehouse, a store executive and aviation buff, commissioned the special flight to advertise his store's annual autumn sale and sold souvenir pieces of the silk to customers at a profit. Parmelee's chilly hour-long, sixty-mile trip to the east, principally following railroad lines between the two Ohio cities, was the first airplane flight that carried commercial freight, but due to cost and the capabilities of airplanes, such flights would not be common for another few decades. The B also served as the inspiration for two smaller airplanes the company wanted to sell to exhibition pilots. Pilots with the company's exhibition department sometimes flew a physically smaller version of the B, the single-seat R (for Roadster) before audiences in Canada and the United States; while the R was significantly shorter in wingspan than the B, it used the same engine as its larger sibling and could achieve higher speeds. Cal Rodgers's *Vin Fiz* EX was also based on the B. It used the same engine as its larger predecessor, while having a wingspan falling between the B and the R. While neither the R nor the EX proved commercially popular, there was consumer demand for the B. In early 1911 the firm claimed that its customers experienced a two-month delay between ordering one and its delivery "on account of the number of orders which the company already had," though its sales ledger suggests that it sold only five airplanes in the first six months of the year. The Model B also piqued the interest of the U.S. military, which wanted to expand its arsenal beyond the single 1909 Wright airplane purchased by the Signal Corps. The War Department acquired three Model Bs for the army and one for the navy. Since it did not want to limit the U.S. fleet to the products of one maker, it also purchased one Burgess-Wright and three Curtiss airplanes for the army and two Curtiss planes for the navy in 1911. The Wright Company did not offer its wares to the War Department at a special price—the Signal Corps bought at retail. Few of these early airplanes had long service lives. The army's Wright Bs fared especially poorly, with only one surviving long enough to be condemned, along with other pusher aircraft, in 1914, and they could be deadly. Lieutenant Lewis Rockwell and Corporal Frank Scott died when Rockwell lost control of Signal Corps no. 4 in a glide at College Park, Maryland, in September 1912, while a squall on Manila Bay, in the Philippines, destroyed Signal Corps no. 7 as the army towed it from a forced landing in August 1913.[17]

The Wright B was not the technological laggard that later Dayton-built models became. It was very similar to the leading airplanes built by other U.S. companies in 1910 and 1911 that were of comparable construction and had similar aeronautical capabilities. Indeed, Burgess deemed the Wright B an aircraft for which there would be such public demand that it signed an agreement with the Wright Company to make exact copies of the B in Marblehead. Soon after learning to fly at the Wright Company's Huffman Prairie school, Harry Atwood landed one of these copies, which were marketed as the Burgess-Wright F, on the lawn of the White House in July 1911. (Atwood did not need to worry about close questioning from the Secret Service; President Taft expected his visit.) The licensing agreement collapsed in 1913, with Orville Wright upset that Burgess workers—now managed by former Wright Company general manager Frank Russell—were making unauthorized improvements to the airplane's structure. The principal Curtiss product then on the market, the single-seat Standard D, was also a wood-frame-and-cloth biplane, though it employed ailerons instead of the Wright wing-warping method of control and used a tricycle landing gear (instead of the two wheels, as on Wright planes). Customers could purchase it with either a four-cylinder ($4,500) or an eight-cylinder ($5,000) engine. But customers buying any of these airplanes had to land them on the ground. Seaplanes were becoming popular, and the Wright Company was slow to recognize the trend.[18]

The first water-based airplanes came from Europe. Henri Fabre's 1910 flights in the first seaplane, *Le Canard*, on the Étang de Berre, a lagoon in the Mediterranean near Marseille, sparked great interest among aviators, as it used bodies of water to take off and land. Rivers, lakes, and harbors were naturally amenable to seaplanes and did not require the same sort of upkeep required of flying fields. Curtiss, trying to avoid additional patent-related lawsuits from the Wright Company, found the new technology especially appealing, as did the Burgess Company, with its yachting heritage. Curtiss established a design standard for seaplanes in 1912 that gained prominence as the Model F in 1913. While the land-based Model D was Curtiss's most successful 1911 airplane, flying boats, and especially the Model F, dominated the company's catalog from 1912 until the outbreak of the First World War. Burgess was slower in moving to an independent seaplane design, but the landing gear of its Model H, introduced in the spring of 1912, could be removed and replaced with floats for water-based operations. The Wright Company took another year, waiting until 1913 before it brought its CH (a Model C with floats instead of wheels) to market. With its license with the Wright Company expiring and with success in selling two of its Model K seaplanes to the U.S. Navy, Burgess was looking for new partnerships in 1913. Late that year it obtained a license from British inventor John Dunne to manufacture seaplanes using his patents, and these seaplanes proved popular. The U.S. War Department purchased several Burgess-Dunne models; the Canadian military's first airplane was also a Burgess-Dunne. The Burgess Company put most of its efforts after 1913 into seaplane development. Both Curtiss and Burgess took advantage of local geography, using Keuka Lake and Massachusetts Bay to develop their flying boats.[19]

Water-based aviation was not a simple proposition in landlocked Dayton in 1912. The Great Miami and Mad Rivers did not provide the Wright Company with the opportunities that the large bodies of water in New York and Massachusetts provided its competitors. Land-based aviation remained the company's principal field, and it attempted to satisfy its potential clientele with the Model C. While Orville Wright believed that the new airplane would "be a dandy" and in high demand, the C proved a disastrous product for the Wright Company's fortunes. Described by Tom Crouch as being obsolete upon exiting the factory when compared with contemporary European models, the "slow, tail-heavy, and unstable" Wright C still attracted the attention of the U.S. Army, which ordered seven of the airplanes to use as scout craft. Slightly smaller than the Model B, the two-seat C used a sixty-horsepower, six-cylinder engine (the B relied on a thirty-five-horsepower, four-cylinder motor) and had four control levers against the two found on the Wright B. Those control levers also affected the Wright Company's fortunes. The continued use of nonintuitive levers by Dayton airplanes was starkly contrasted by the more intuitive stick-and-pedal method of control then becoming common on such French aircraft as those built by Blériot and Robert Esnault-Pelterie and, soon, by Curtiss. The C's greater engine power in an airframe slightly smaller and weaker than that of the B led to significant safety and reliability problems for the Wright Company. Still, Orville remained confident of the airplane's prowess in the face of its poor publicity and the new models introduced by competitors. He and then-manager Grover Loening even reduced the power of the C, offering a version with a four-cylinder engine for sale for $5,000 (the six-cylinder Model C sold for $6,000) on the company's December 1913 price list. The U.S. military's experiences with the C, though, resulted in there being few customers for Orville's dandy.[20]

Safety problems beset the Cs operated by the U.S. Army's Signal Corps, whose pilots found flying it dangerous to life and limb. Of the seven Cs acquired by the army in 1912 and 1913, accidents destroyed six by the end of 1914 and killed seven pilots and passengers, including Wright Company pilot Arthur Welsh. Two soldiers also died in a Wright B crash in 1912, and another died when his Burgess-Wright version of the C (the J Scout) crashed in 1913. The safety record of the army's Curtiss pushers was somewhat better, as only three soldiers met their ends in an airplane from Hammondsport, but the Signal Corps flew fewer Curtiss airplanes than Wright models in its first years. Regardless of what one flew, military aviation was a dangerous profession. Of the fifty-three individuals who flew for the Signal Corps between 1907 and 1914, fourteen (26 percent) died in airplane accidents, and several more sustained injuries. Grover Loening, who went to work as an aeronautical engineer for the army in San Diego in the summer of 1914, after leaving the Wright Company, noted that the army pilots whom he worked with—many of whom had trained at Huffman Prairie—encountered problems when flying the Wright Company's airplanes. Their travails led him to believe that the Model C, which the company first delivered to the army before Loening started working in Dayton, "was too tail heavy and didn't answer the controls properly." He also believed that the "the pusher planes were not only easily stallable, but in a crash the engines generally fell on and crushed the pilots." The one company executive whose opinion counted, however, disagreed. Orville Wright was convinced that the C, with its more powerful engine, represented an improvement on the B and was no more dangerous than any contemporary airplane, and he rebuffed such assertions and attributed the C's many accidents to pilot errors. Wright's belief that the C was a safe plane when flown by a supposedly capable pilot carried little weight in Washington.

The army's chief signal officer, Lt. Col. Samuel Reber, condemned the one flyable C and all other pusher airplanes still in army service, including those made by Curtiss and two Wright Model Ds, in February 1914. The army then issued new standards, requiring that all future airplanes added to its arsenal employ tractor configurations, with the engines and propellers to the front of the pilot. All the Wright Company's planes then in existence used the pusher configuration, and the company built only pushers as long as Wright was president. While the Curtiss company quickly shifted to producing planes that satisfied the new army requirements—its J and N series tractors—the Wright Company was uninterested in making the change. The company that provided the country's first military airplanes was never again a significant supplier of aircraft to the U.S. Army or Navy.[21]

The loss of the U.S. military as a customer was a result of two years of corporate decline that occurred immediately after the loss of the company's first president. At the beginning of May 1912, Wilbur Wright returned to Dayton from a trip to Boston, where he had been consulting with Frederick P. Fish of Fish, Richardson, Herrick and Neave, one of the company's patent attorneys, about the status of the case against Glenn Curtiss then before U.S. District Judge John R. Hazel in Buffalo. He came home unwell, perhaps from consuming bad shellfish (oysters, especially, were a favorite of both brothers) while in New England, though he was not sure. Typhoid fever is now treated with antibiotics and is rarely fatal if treatment is available, but the Wrights lived before penicillin. In 1896, as a young bicycle mechanic and printer, Orville Wright had caught typhoid fever himself, perhaps from water from a contaminated well behind the brothers' bicycle shop, but he fought off the sickness and slowly recovered. His brother would not be so fortunate. At first, Wilbur's case did not seem critical, but he worsened as the month progressed. On 17 May, Katharine wrote a letter to Reuchlin, the eldest Wright sibling, in Kansas, urging him to come east as quickly as possible. The next day, Milton Wright recorded, the feverish Wilbur had "an attack mentally, for the worse. It was a bad spell. He is put under opiates. He is unconscious mostly." Wilbur drifted in and out of consciousness over the next week, his health continuing to deteriorate, before he finally died early on 30 May. He was forty-five years old. "A short life," wrote his father, "full of consequences." As word of his death reached the Dayton press and the news wires, condolence calls and telegrams arrived at 7 Hawthorn Street from around the world—including one from Glenn Curtiss. Starling Burgess told *Fly Magazine* that Wilbur was "more responsible for the advancement in aviation than any who have been identified with it." Andrew Freedman called him "a most remarkable man in every respect" who was "able on almost every subject, even those which were not related to his life study." President Taft stated that he "deserves to stand with Fulton, Stephenson and Bell" in the pantheon of inventors. Undertakers from Dayton's Boyer Mortuary soon arrived at the family home and embalmed Wilbur in his childhood bedroom, preparing him for burial.[22]

But where to have the service? Though the sons of a prominent United Brethren in Christ bishop, one who still sometimes made note of his scripture readings in his spare diary, neither Wilbur nor Orville was a member of a church—United Brethren or otherwise. Given the fame of the deceased, the seating capacity of a church's nave was paramount, and downtown Dayton's First Presbyterian Church fit the bill. After a packed twenty-minute service conducted by the Rev. Dr. Maurice E. Wilson, Wilbur was buried on a wooded hill in Woodland Cemetery, Dayton's most prominent burial ground, on 1 June, a Saturday. A wreath from the company was among the many flowers covering his grave. Dayton industry ceased for the service; at three thirty workers lay down tools, church bells tolled, streetcars stopped in the middle of their routes, and telephone operators did not connect calls for several minutes. Company directors Russell Alger and Robert Collier, as well as NCR owner John H. Patterson and U.S. Representative James H. Cox, the owner of the *Dayton Daily News*, were among the honorary pallbearers, and Frank H. Russell attended from Marblehead. But business would go on amid the mourning. The factory reopened for work on 3 June, the same day that lawyer Ezra Kuhns read Wilbur's will to the family. He left an estate of nearly $280,000, with most of the cash going to Reuchlin, Lorin, and Katharine, and his shares of any patents and company stock to Orville.[23]

As Wilbur lay in Dayton on the verge of death at the end of May, Orville sent identical telegrams to several company officers, including Russell Alger, Andrew Freedman, and lawyer Pliny Williamson. He told these men to prepare for a new president. "Wilbur gradually sinking. Doctors have no hope." Wright's telegram served as a metaphor for the subsequent fate of his company. Its failure partly resulted from the brothers' neglect of the company's products in favor of the courtroom and of pilot education, as well as more personal matters. Before his death that May, Wright spent significant time pursuing the patent infringement lawsuit against Curtiss, providing testimony in Dayton and New York. He had little time to devote to the factory or the drawing board in the four months before he fell ill. Orville Wright spent the months before his brother's death pursuing royalties from exhibition aviators (to try to get pilots to fly Wright airplanes, the company discounted royalty fees 75 percent until the resolution of the patent lawsuit against Curtiss) and corresponding with several aviators connected with the Wright Company, including Clifford Turpin, Roy Knabenshue, Albert Merrill, and Signal Corps Aeronautical Division commander Capt. Charles Chandler, on technical aviation matters. After his brother's death, Orville spent much of the summer keeping their grieving father company and working with the architects and builders responsible for the construction of the new Wright estate, Hawthorn Hill, in suburban Oakwood. Orville's assumption of the Wright Company presidency and Wilbur's role in the patent lawsuit also removed him from opportunities to spend significant time at the drawing board refining his company's aircraft, making it difficult for the company to respond effectively to the problems pilots encountered with the C.[24]

In the weeks after Wilbur's death, Orville's mind was not immediately on business. He went on several long drives around the Dayton area with his bereaved father, visiting the lot in Oakwood where construction would soon start on the Hawthorn Hill estate (to which he would move in 1914 with Milton and Katharine), and going to Springboro, to the south, and West Liberty, to the north. Meanwhile, Milton remained at home, responding to the letters of condolence arriving at 7 Hawthorn Street. To a former student living in Oregon, he wrote that Wilbur "has gone to be among the angels of God." The late company president now "soared victorious as a conqueror to his throne" and "we are left to weep." The Wright Company board of directors, meeting in special session, passed a resolution noting that with Wilbur's death "this company has suffered the irreparable loss of its chief inspiration and guiding spirit, and the Directors have lost the personal association of a man of the highest character and possessed of the loftiest business and scientific ideal." The new company president had very little time for his personal laments, as he was quickly forced to grieve anew. Former exhibition department aviator Al Welsh, in College Park, Maryland, to fly a Model C recently purchased by the Signal Corps, lost control of the airplane and crashed, killing himself and his passenger, Lt. Leighton Hazelhurst, Jr. With Katharine at his side, Orville quickly traveled to Washington, D.C., for the funerals of the Kyiv-born Welsh at the Adas Israel Congregation, and of Hazelhurst at Arlington National Cemetery. Alpheus Barnes and factory superintendent Arthur Gaible also represented the company at the services. Expressing a tinge of anti-Semitism, common at the time, Katharine Wright wrote that Welsh's family, which attended a Conservative, not Orthodox, synagogue, showed "their Jewish traits much more than" he did.[25]

The power and structural integrity problems that affected the C and led to the deaths of Welsh and Hazelhurst also affected the Wright Model D, of 1912, which the company offered with either a four- ($5,000) or six-cylinder ($6,000) engine. Built as a quick-climbing single-seat military scout plane closely resembling the Model R, the Model D was, in Orville Wright's estimation, "the easiest to control of any [airplane] we have ever built," even at the relatively high flight speed of sixty-five miles per hour (fifty to sixty miles per hour being more common speeds). Wanting to give the Wright Company another chance, the Signal Corps purchased two of the aircraft, and in advertisements the Wright Company boasted that one of those Ds set an American record in its proving tests at College Park, having "consistently demonstrated a climbing of 1640 feet in 3 minutes." Nevertheless, the Signal Corps' pilots were less impressed with its operability than was Wright. Lt. Hap Arnold cautioned Wright in March 1913 that the army was unlikely to acquire more airframes. While the corps saw the value of an airplane with the general characteristics of the D, Arnold noted, only Lt. Thomas D. Milling "was capable of flying" the D itself. While counseling Arnold that the D's "high speed in landing is its only draw-back," Wright continued to assert that the model was "the easiest machine we build" to fly. Arnold proved prescient. The army did not order additional aircraft, and its two Ds remained in its inventory for a year before being condemned by the army and discarded with the corps' other pushers. Aside from these two airframes, it is unclear how many other Ds workers at the Dayton factory built, if any. The difficulty in flying and the disturbing safety record of the army's Cs and Ds, and, to a lesser extent, of its Curtiss and Burgess pushers, led to their ultimate condemnation.[26]

Wright Company airplanes never found extensive use by the pilots of the U.S. Navy. The fleet's preference for Curtiss models led to the Hammondsport builder being nicknamed the Father of Naval Aviation. The navy purchased more than two hundred Curtiss Model F flying boats during the 1910s. The Wright Company was aware of the expanding seaplane market, and finally made an effort to enter it in 1913. It found that year, though, to be a poor one for sales—and a difficult year to be based in Dayton, though perhaps one where seaplanes would have been more useful locally. While the company attempted to respond to the interest in seaplanes with the CH and G pushers and place them and the land-based E and F models before consumers, it also had to contend with the extensive damage caused to the Miami Valley's infrastructure by the Great Dayton Flood that March, when three successive storms dropped approximately ten inches of rain on frozen ground within three days. Unable to soak into the soil, the rainwater ran into the Great Miami River and its tributaries and caused them to overflow, bursting levees and drowning downtown Dayton in up to twenty feet of water. With its factory on slightly higher ground more than two miles west of downtown and the Miami River, the Wright Company's shop remained dry, though the first floors of Orville Wright's home and office in west Dayton were inundated (where the glass plate negatives of photographs of the brothers' aeronautical work, including the famous 1903 photograph of the first flight, were stored). Water flooded more than fifteen thousand residences throughout the city, including those of company employees Harry Harrold, Daniel Farrell, and Frank Quinn. Ruptured natural gas lines caused fires, many of which were inaccessible to firefighters due to the high water, and the fires destroyed blocks of the city. The flood spared National Cash Register's complex, located on higher ground in south Dayton, and NCR owner John H. Patterson coordinated the emergency response to the rising waters after Mayor Edward Phillips and other city officials became all but invisible, with City Hall inundated downtown. Instead of building cash registers, NCR carpenters used the company's woodshops to build small boats to enable rescuers to reach stranded people reduced to escaping from their attics through holes in their homes' roofs, while NCR cafeterias served hot meals to nearly five thousand displaced people. By 28 March, Dayton was physically dry, but the damage throughout the city was soon clear. Almost twelve hundred horses had drowned and debris and muck covered the streets. Many homes were underwater (people can still find traces of 1913 mud behind lathwork in older homes), and centrally located factories and businesses of the city's downtown industrial base took a major hit. The Barney and Smith Car Company, a builder of high-end railroad passenger cars and the employer of thousands, suffered more than a million dollars in damages at its downtown plant and went into a receivership from which it never recovered. Fire destroyed the downtown Rike-Kumler department store. And, of course, there was a human cost. The flood directly caused 361 deaths in Dayton and over $100 million (in 1913 dollars) in general property damage. It also led to the creation of the Miami Conservancy District, which built a series of dry dams on local waterways for flood control. One of these dams, on the Mad River, would protect Huffman Prairie Flying Field from future development by the U.S. Air Force as Wright-Patterson Air Force Base grew, by placing the field within the river's potential floodplain and making it an inappropriate site for the construction of permanent structures.[27]

FIGURE 6.2. Fires from broken natural gas lines were a major problem after the Great Dayton Flood of 1913, as shown in this photograph of burned buildings in west Dayton. *Courtesy of the Dayton Metro Library, 1913 Flood Collection*

While the Wright Company's factory infrastructure survived the flood, the devastation in Greater Dayton "caused a loss of almost a month's time" in its production schedule. Orville found it "almost impossible to get our men back to work" and, due to the lack of functioning streetcars, the company was "compelled to run short hours in order to allow [workers] time to get to and from their homes." Though records do not suggest that the company made special financial allowances directly to its employees who were affected by the flood, it donated $5,000 (nearly $114,000 in 2010 dollars) to the Citizens' Relief Committee. The company then turned inward. Reflecting his firm's limited size and stature in the Greater Dayton business community, as well as his own highly introverted personality, Orville Wright was not among the local corporate executives, led by NCR's John Patterson, who served on the relief committee and organized the city's recovery efforts, though he personally donated more than $1,250 to recovery efforts in the three years after the deluge. For the Wright Company in the summer of 1913, the goal was returning to business as usual, not the rebuilding of Dayton.[28]

But Dayton did rebuild. Orville Wright's enterprise came through the flood in better physical and financial shape than did Barney and Smith, but it did not benefit from the "gigantic boom" in the construction and building trades that bolstered the city's economy in the immediate aftermath of the flood. Indeed, as a company in, but not of, Dayton, it was unlikely to do so. Dayton residents were not its customers, and its facilities were too small to have any significant role in the reconstruction of the Gem City. Instead, the company's workers returned to what they did best: building airplanes, though they undoubtedly resumed work stressed from the travails of flood recovery, with homes and lives under repair. Once activity recommenced on the factory floor, workers turned to building the four new Wright Company models of 1913. These biplanes, which reflected the traditional Wright pusher style, struggled to find buyers. The principal land-based model of that year was the $5,000 Model E, a smaller and lighter version of the Model C intended for exhibition flying. Introduced that autumn, it was the only Wright Company airplane built during Orville Wright's ownership of the company that used a single pusher propeller; all other Wright models had two. During tests at Huffman Prairie Flying Field, Wright Company workers were able to dismantle an E into its three component sections and prepare it for shipment in twelve minutes, and the company noted in advertisements that the model was "designed particularly for ease in assembling and taking down." A determined Orville Wright (who no longer had a stable of exhibition pilots who could put a new model through its paces) conducted most of the airplane's flight tests himself, and he personally flew an E to demonstrate his automatic stabilizer—which won him the Aero Club of America Trophy in 1913—in the chill of that December in Ohio, flying seven times around Huffman Prairie with his hands free of the control levers. Historian John Carver Edwards asserts that the E was "quite popular on exhibition tours because of the speed and ease associated with its assembly and disassembly," and the newly formed Elton Aviation Company of Youngstown, Ohio, noted that it purchased a Model E in the autumn of 1914. However, few other instances of others flying the airplane are mentioned in the aeronautical press (which generally noted the make of an aviator's airplane but rarely recorded the specific model), and company records are silent on how many were sold.[29]

Another of the four 1913 airplane models was ordered by the U.S. Army. The two-seat Model F, ordered in March for $9,500, was the last airplane Wright Company workers built according to army specifications. Principally designed by Grover Loening and known by army aviators as the Tin Cow, the F was the first Wright Company model with a fuselage, a more aerodynamic feature that French aviator Louis Blériot, among others, had first employed on his airplanes nearly five years earlier. After preacquisition flight tests that

FIGURE 6.3. A Wright Company Model H flying at Huffman Prairie Flying Field, 1914. Orville Wright may be the pilot. *Courtesy of the Library of Congress*

resulted in what was originally a tractor airplane design being built as a pusher, the army added the Model F to its inventory in June 1914. This one airplane, the only F ever built, had a very short military career, though, completing only seven flights before the army discarded it in June 1915 after the condemnation of all pusher models. Lt. Herbert Dargue, the only Signal Corps aviator to fly the F, found it "uncontrollable on the ground and out of date." The influence of the F could be seen in the slightly smaller Model H of 1914—another Wright Company airplane with a fuselage—if one could find a Model H to see.[30]

While Wilbur and Orville Wright personally experimented on hydroplanes and floats on the Miami River in early 1907, their company was a late arrival to the commercial seaplane market. Its first attempt to enter that arena was the two-seat CH (Model C, Hydroplane; also written C-H), a C modified with pontoons attached to its undercarriage instead of wheels. Orville Wright himself was the pilot for most of the aircraft's test flights from the Miami ("over one hundred," it was claimed); he even took the curious aloft, doing "a large business in carrying passengers, taking one up after another, often despite winds of as high as 10 to 15 miles an hour." Grover Loening, who arrived in Dayton to work for the Wright Company just as the CH entered the market, advised the aviation community that the CH was an airplane of "remarkable efficiency and air-worthiness," one that Wright designed to be "far more air-worthy than any machine of this type that has yet been built." Wright designed the $6,125 CH for the consumer market in an era of poor intercity roads. (In 1919, Lt. Col. Dwight D. Eisenhower, an observer on the army's Trans-Continental Motor Truck Trip, would note that after his convoy reached Illinois, "practically no more pavement was encountered until reaching California.") Loening wrote that the CH would fly "between towns and to [open] up inaccessible country over the thousands of shallow streams that are open to no other kind of navigation." The company asserted that the airplane "shows higher efficiency than has ever been attained in marine flying," but independent pilots found that it flew poorly, as the pontoons (in various arrangements) increased drag and made it more difficult for a pilot to turn the airplane. Poor flight was a significant strike against an airplane based on another Wright Company model, the C, which had developed a reputation as an unsafe aircraft. Few of the CHs sold; while the company's ledger is silent on the matter, used-airplane advertisements in the aeronautical press provide no evidence of private pilots attempting to unload an unwanted CH. The U.S. Navy purchased three of the planes but found them to be marginal performers. Curtiss flying boats with hulls dominated the market for airplanes that operated from water, and the Wright Company did not pursue development of the CH, turning instead to its own version of the flying boat, a more robust airplane with a fuselage instead of open seats on the lower wing.[31]

That model was the G, a seaplane principally designed by Grover Loening and tested by Loening, Orville Wright, and Oscar Brindley on the Miami River. It would be the Wright Company's answer to the Curtiss Model F. The company's advertising claimed that the two-seat G—slightly more expensive than the CH, with a list price of $7,000—was "the development of years of careful experiment," though Loening had arrived in Dayton in the summer of 1913 and the company announced the G's release that autumn, allowing little time for Loening to take an airplane from the drawing board to the air or to incorporate features specific to his own 1911 and 1912 designs. Richard Hallion has also cast doubt on the G's supposedly lengthy development, writing that the essence of the airplane had changed little; its hull was "joined to an essentially Model C airframe." At the time, the company asserted that the Aeroboat was "perfectly adapted to the use of sportsmen for safe and comfortable travel over water at high speed," high speed then being sixty miles per hour, and Wright expected it to be used by the post office to carry mail in remote "regions of the west, in Alaska, and along the coast," according to

Loening. Writing decades later, Loening was more measured in his assessment, noting that those years of careful experiment had one overarching goal: not following the lead of Glenn Curtiss. He stated that Wright "would never have approved of the . . . Wright Model G . . . if its appearance had not been so totally different from the Curtiss F boat that was sweeping the field" and remained in production, with various modifications, from 1913 to 1919. The Curtiss F and the Wright G used dramatically different noses and tails; the Wright model was slightly larger (though lighter) than the Curtiss and used a sixty-horsepower six-cylinder engine against the one-hundred-horsepower V-8 of the Curtiss. Later production models of the G were the first of the Wright Company's airplanes to adopt more mainstream airplane controls, abandoning the complicated control levers used by earlier Wright airplanes for a wheel-based control, one that it now considered "more suitable for flying" long distances. *Flight's* editors applauded the change, writing that the new wheel, similar to the control wheels long used on Curtiss airplanes, was "an improvement on the old type Wright control . . . one which will make a strong appeal to pilots." Neither the introduction of the control wheel or of the Model G, though, was enough to return the Wright Company to a prominent place in the North American skies.[32]

FIGURE 6.4. A Wright Company Model G, or flying boat, in flight. *Courtesy of the Library of Congress*

While the release of the G received a fair amount of attention in the New York newspapers and in aeronautical journals, it received less attention from customer checkbooks. The U.S. Navy bought the first production model, with which Harry Atwood conducted test flights on Lake Erie near Toledo in May 1914, but after Lt. Louis H. Maxfield arrived in Dayton to inspect the Wright G, in the summer of 1914—shortly after Loening left for San Diego and his new job with the army—the navy soured on the Dayton product. Maxfield, who later served as an airship commander (and died in the 1921 crash of the ZR-2/R38 over Hull, in England), did not recommend that the navy, which was already a major Curtiss customer, add the Model G to its arsenal.[33] The navy, more concerned with strategy connected with the opening of the Panama Canal that year, followed his advice. Atwood was more positive, acquiring for himself the second Wright G produced. Capt. J. William Hazelton of the First Provisional Aero Squadron of the New York National Guard may have personally acquired another G. Hazelton wanted to let the U.S. Army use his "Wright-type flying boat" (which may, indeed, have been an airplane from a different maker, or a home-built plane, only styled after the G) in support of Gen. Venustiano Carranza's forces in the Mexican Revolution. The army did not take him up on his offer, and Hazelton soon advertised his airplane for sale. Ernest Hall, a young aviator from eastern Ohio, also acquired a G, flying it at Conneaut Lake, in western Pennsylvania; his airplane, reduced to a partial skeleton, later became part of the collection of the Smithsonian's National Air and Space Museum. But, as with other Wright Company airplanes, no specific figures documenting the number of Wright G aircraft sold to the public or the military are extant. The company's accounting entries for 1913 suggest that all sales brought in slightly more than $22,000 that year, a figure not indicative of great demand for Wright Company airplanes, which cost between $5,000 and $7,000 each.[34]

Before the summer of 1914, when the U.S. Army grounded all pusher airplanes, the Wright Company remained hopeful that the military would remain a customer. Indeed, the 1914 and 1915 models coming out of the Dayton factory show that the company viewed the military, and not exhibition aviators, as its principal customer, with the catalogue containing models built specifically for military uses as war brewed in Europe. Unfortunately for the company, the military remembered the poor performance of its earlier Wright aircraft and, insisting that its hangars house only tractor-style airplanes, did not return the attention. Wright company products were technologically obsolescent, of little value for the nation's arsenal. Attention for Wright Company products was hard to obtain. *Aeronautics, Flying,* and *Aircraft,* the major U.S. trade journals, all passed on providing their readers editorial comment on new Wright models, though *Flying* did run advertisements for the new HS. Instead, the Wright Company that appeared in their pages was a victorious patent litigant or a potential acquisition by other companies; it was not viewed as a manufacturer of significance, even though it still employed people who built airplanes.

Some workers—how many is unknown—remained employed at the Dayton factory, building the Models H (1914) and HS (1915) to be military scouting airplanes also capable of carrying bombs. Like the Model F, the H was essentially a land-based Model G, with wheels. The HS was a smaller and lighter version of the H, built because Orville Wright thought that the H was too slow in the air. Both airplanes included a canvas-covered fuselage, continuous from the cockpit to the airplane's tail, instead of one made of wood and metal. What press attention these two airplanes attracted came predominantly from Britain, not a country where the U.S. Wright Company marketed its wares. There, the H and HS received attention for what they were not: significantly different from earlier Wright biplanes. The Royal Aeronautical Society's *Flight* wrote of the HS that "it is remarkable how the main [Wright] characteristics have been adhered to . . . it is still the Wright biplane of old." Smaller and faster than the H, the HS, promoted as a military flyer, was the final Wright Company airplane with a double vertical rudder and pusher propellers. The U.S. military, lacking confidence in the Wright Company and in the U.S. aircraft industry generally, acquired neither airplane. It ignored the Dayton products in the same manner that the Wright Company ignored the standards the U.S. military required after its condemnation of pusher aircraft. But a tentative production lifeline for the Wright Company came from Europe, where the U.S. Army's standards did not apply. With its demand for military aircraft outrunning the supply available domestically as it fought on the Continent and protected the home islands, the British military expressed interest in acquiring the Wright H. The Royal Flying Corps placed an order for twenty-five aircraft, and the Wright Company prepared an approval model to send across the Atlantic to seal the deal. Quality control, it turned out, was suffering with the rest of the Wright Company. Solder found in the fuel pump of the approval model kept that H on the ground and caused the RFC to cancel its entire order. No Wright Company planes flew in the First World War. After the United States entered the fray (more than a year after the Wright Company left Dayton), army pilots on active service flew U.S.-manufactured copies of the British DH-4, designed by Geoffrey de Havilland, while naval aviators flew the DH-4 and several other Curtiss and European models. Some Model H or HS airplanes may have found a home in Mexico with the revolutionary forces of Francisco ("Pancho") Villa; in April 1915, *Aerial Age Weekly* noted that he "had just acquired six new Wright machines."[35]

The U.S. Navy, though never a major purchaser of Wright airplanes, had not soured on the Dayton company as thoroughly as had the army. In the spring of 1915, two months before the German sinking of the Cunard liner *Lusitania* off the coast of Ireland helped turn U.S. public opinion against Kaiser Wilhelm II, the navy opened bids from fourteen airplane makers, large and small, for nine hydroplanes. In 1913 naval aviator and designer Holden C. Richardson wrote several letters to Orville Wright concerning the development of the Model G, and he now traveled to Dayton to encourage Wright to have his company enter a bid. The Wright Company, it turned out, was one of the successful bidders, promising to deliver a new airplane built to the bid's specifications to the U.S. Navy Aeronautic Station, in Pensacola, by 15 June for demonstration trials, for a price between $5,200 and $9,740. The airplane that the navy purchased that fall after its trial was the last Wright Company airplane to enter U.S. military service, the Model K seaplane. It was also Orville Wright's final attempt at building a commercially successful airplane. With the navy's standards limiting his design options, Wright was forced to break from his traditional pusher biplane form and from his devotion to wing warping. The navy made the K the first tractor airplane built by the Wright Company and the first Wright machine to employ ailerons instead of the wing-warping mechanism that had controlled all Wright aircraft since the brothers' 1899 glider. Orville's last gasp was just that; the K, though appearing "trim and purposeful" in photographs examined years later by Richard Hallion, fared poorly commercially, with the only recorded sale being that of the airplane sent to Pensacola, and the aeronautical press ignored it. With the failure of the K, Orville Wright left the world of aviation design. Within five months, he would leave the aviation industry.[36]

The workers at the factory would continue to build even after Orville severed his ties with his namesake, but only for a short time. The last airplane built by Wright Company workers in Dayton before the company merged with Glenn Martin's firm to become Wright-Martin (later, Wright Aeronautical) and relocated to New Jersey, the Model L, was marketed as a single-seat tractor military scout and bore little resemblance to the pusher airplanes favored by Orville Wright. It used a direct-drive engine (not one driven by chains, which harked back to the Wrights' bicycle days), ailerons, and a wheeled undercarriage. It was also inherently stable, unlike earlier Wright airplanes. Hallion described it as an aircraft with "an almost 'amateur' or model airplane-like appearance," one that "obviously lagged behind the contemporary design standard of combat aircraft on the Western Front, and, not surprisingly, failed to secure production orders." The U.S. Army ignored the airplane's release, as did the U.S. aeronautical journals. The L received a tepid review in *Flight*, which noted that the airplane's "lines do not impress one as being particularly pleasing, and it would appear to have been quite possible to have improved these considerably without necessarily increasing the cost. However, according to reports the machine flies very well, and after all that is the main consideration." The L marked the end of the line for Wright Company production at the Dayton factory, and for the building of original, company-designed airplanes.[37]

The Dayton-Wright Airplane Company would use Orville Wright's name as it designed and built airplanes in the Dayton area during the First World War, and the Wright name would fly alone on Wright Aeronautical models made between 1919 and 1929, but Orville Wright would not be involved in either company's management. Wilbur and Orville Wright's attempt to conquer the airplane market as they had conquered the skies at Kill Devil Hills, which began with the assistance of some of the titans of finance and industry in 1909, ended quietly with the L in 1916.

# 7

## Turning Buyer Attention the Company Way

*Advertising*

The Wright Company's lack of commercial success was not the result of the sort of secrecy that the brothers demanded in the years before 1908. Hundreds of thousands of people saw their flights in New York and in Europe, and newspapers closely covered their professional lives, providing plenty of unpaid publicity. But the brothers had no control over what newspaper reporters wrote. They realized that they needed to use paid advertisements if they were to ensure that their products were placed before consumers' eyes in exactly the way they preferred. Wilbur and Orville Wright were certainly not naive when it came to advertising their previous businesses. They advertised their printing and bicycle sales and repair businesses in the Dayton press, and even printed their own newspaper-cum-advertising circular, *Snap-Shots at Current Events*, to publicize the Wright Cycle Company from 1894 to 1896. In the competitive arena of early aviation, where their 1906 patent was often ignored or denigrated, they realized that their company needed to turn the eyes of potential aviators away from articles about their propensity to sue for patent infringement or from writings about the achievements of airplanes made by Curtiss, Burgess, or European companies and toward the craftsmanship of their workers in Dayton. The company's publicity budget—an informal one that never gained its own ledger listing—went to purchasing space in the aeronautical press. Aviation was too expensive a pastime for all but the wealthy, and advertisements placed in general-circulation newspapers or magazines would have been a waste of money. The closest the company came to advertising in a general-circulation publication were its placements in "The Air Scout," a supplement sponsored by the (civilian) U.S. Aeronautical Reserve that ran in the high-society magazine *Town and Country* in 1911. *Town and Country*'s wealthy, social register–climbing readers were the sorts of private individuals who could afford to buy an airplane. But advertisements in aeronautical periodicals reached more actual pilots than the *Town and Country* ads, which were mostly read by armchair aviators.[1]

Several periodicals attuned to the aviation community appeared in the years around 1910. Ernest Jones's *Aeronautics*, first published as the *American Magazine of Aeronautics* in 1907, was one of the earliest. Until its closure, in 1915, *Aeronautics*, usually a monthly, served as the official journal for the Aeronautical Society of America. It faced stiff competition from *Aircraft*, a magazine first published in 1910 by the eccentric Alfred W. Lawson, who claimed coinage of the term *aircraft*, promoted populist economic ideas during the Great Depression, and who in the last decades of his life developed a system of physics and philosophy, Lawsonomy, that critics would label a cult. Lawson asserted that *Aircraft* had a monthly circulation of fourteen thousand by 1911, though the figure is impossible to verify. However many readers they reached, early company advertisements in these and other trade journals placed before Wilbur Wright's death, in May 1912, emphasized the fame of the Wrights and the records set by pilots flying the company's airplanes. First placed in *Aeronautics* while the company was still renting space from Speedwell, the half-page textual pieces noted that the Wright Company's workers built the engines and airframes of the airplanes the company sold; they were not assembling kits actually designed by other manufacturers. As the Wright Company moved

into its new quarters, in west Dayton, its *Aeronautics* publicity announced the
safety, comfort, and practicality of the Wright B, noting that the pilots flying
the Dayton product beat aviators in Curtiss, Blériot, and Farman airplanes
"in duration, distance, altitude, accuracy of landing, slow flight" and took "the
Hammond Cup for bomb throwing" at the "recent" Harvard-Boston meet in
Massachusetts (in September), a meet sponsored by the Harvard Aeronautical
Society and the *Boston Globe*. Such a company was a stable company, one with
a long-term future. The company did not change its ads frequently; the piece
announcing the Harvard-Boston results as being recent ran from November
1910 to September 1911, twelve months after the event in focus. By May 1911
the Wright Company's *Aeronautics* ad suggested that its products were "not
experiments," unlike many of the airplanes individuals built in their garages.
It also claimed that aviators who wanted to fly their new Wright airplane
that summer needed to order immediately (to "Department A"; inquiries in
*Aircraft* were directed to "Department B") as "already our orders exceed our
production for April and May." But company workers could build only four

FIGURE 7.1. *A typical early Wright Company advertisement from a 1910 issue of*
Aeronautics

airplanes per month at most, and some of those planes were destined for the exhibition department. It would not have taken many responses to the ads for the "peerless machine" that held "all American records" to overwhelm the Wright Company factory, but the ads were not that effective. To try to make readers' eyes linger, the company began including a photograph featuring a Model B on the ground, with two others above it in flight, to *Aeronautics* ads starting that June.[2]

Company advertisements featured the airplanes of the day or the flying school. Exhibition appearances were local matters, their advertisement the responsibility of whatever body organized a fair or festival, and the company did not believe that those individuals perused the aviation magazines. But by 1912 the exhibition department was dead, leaving the point of advertising it in *Aeronautics* or *Aircraft* moot. Only one Wright Company ad (repeated throughout the year) appeared in *Aeronautics* in 1912. Lauding the now-superseded Wright B for its "efficiency and reliability," the company claimed that buyers could equip their new C or EX with "automatic control, silent motors, and hydroplanes." The ad also noted that it was in an EX that Cal Rodgers flew from New York to California the previous autumn, not mentioning his many crashes or that it was an essentially rebuilt airplane that finally made it to the Pacific. For the first time in *Aeronautics*, the Wright School of Aviation made an appearance, with the company providing a partial list of graduates, men who were the "most famous flyers in America." The Wright Company positioned itself as providing the best education for budding pilots hoping to earn a living in the air. It even enjoyed enough of a positive image that other firms referenced it in their own advertisements. Twice in 1912, B. F. Goodrich advertised its Lumina airplane cloth beside the Wright Company's ad, bragging that "the Wright Company and other discriminating aeroplane manufacturers" used the cloth on their airplanes. But that attention was fleeting, as is especially evident in the advertisements that the Wright Company placed after 1912.[3]

The Wright Company's advertisements for its 1913–16 models were more restrained than those of its first years, suggesting that the company realized that its products did not lead the field and that buying a half-page in *Aeronautics* was not the best way to reclaim it. During the spring and summer of 1913, *Aeronautics* and *Aircraft* ran advertisements noting the availability of the E and EX models, with the company advising pilots that Wright airplanes were easy to break down and reassemble for shipping by rail the long distances between appearances, but the ads did not plug the airborne abilities of the airplanes. The same ads, though, did claim an American record for the Model

The Wright Company

(The Wright Patents)

THE NEW WRIGHT
AEROPLANES

For sport, exhibition or military use, over land or water now embody the improvements that have been suggested by the experiments quietly conducted during the past ten years.

THE WRIGHT FLYING
SCHOOL

Located at Dayton opens May 1st, for the season of 1915. Tuition $250. No other charges of any kind Enroll now. Booklet on request.

The Wright Company

DAYTON, OHIO          New York Office: 11 Pine St.

FIGURE 7.2. A typical late Wright Company advertisement from the 30 April 1915 issue of *Aeronautics*, showing a Model HS over Huffman Prairie

D's ability to climb 1,640 feet in three minutes. And by that fall, the company wanted readers to be aware of its belated entry into the seaplane arena. Grover Loening, especially, made several attempts to promote the Wright Company's products during his tenure in Dayton, in both purchased advertisements and articles on the CH, E, and G that he wrote for *Aircraft* and *Aeronautics* on the invitation of their editors. Once Loening left Ohio in the summer of 1914, the company's ads were of a more general nature, usually avoiding mention of specific aircraft models, and the part-news, part-ad articles Loening wrote on behalf of the company's new airplane models became rare. The company was left with no prominent manager or executive willing to support extensive marketing or to make the effort himself.[4]

In the minds of its other leaders after Wilbur Wright's death, the Wright Company was part of a refined industry where a company's advertising should resemble a nineteenth-century presidential campaign: self-promotion was to be avoided, if possible. Indeed, Alpheus Barnes believed that the airplane market was a limited one, and that airplanes should not be advertised "like breakfast foods." Company ads steered clear of the appeals of a Post or Kellogg (or of a later Eisenhower or Obama) and made no attempt to broaden the appeal of the company's products. Instead, they became akin to the post-1915 Orville Wright, who was intensely private and introverted and avoided public appearances if at all possible. The ads ignored the products of competitors and merely referenced the availability of the company's "various types" of airplanes, or announced to the rest of the industry that the Wright Company was healthy and profitable. One wonders if gossip in the clubhouse of the Aero Club of America concerned a declining company, as one ad took pains to assert that the "season of 1914 [would] be a prosperous one" for Orville and his Dayton employees. Placed after the Wright Company won the first round of its patent infringement lawsuit against Glenn Curtiss, the advertisement advised aviators that international industry-leading builders Curtiss, Farman, and Blériot had infringed on Wright patents. Anyone who decided to fly one of their airplanes and not a Wright Company model would also be an infringer, and risked receiving a subpoena of their own. Buy Wright airplanes and avoid court, they proclaimed. But veiled threats ebbed as 1914 progressed and Orville bought out most of the other directors. By 1915 ads rarely gave more than general endorsements of Wright airplanes, which "now embod[ied] the improvements that have been suggested by the experiments conducted during the past ten years."[5]

Recognizing that the company was spending money ineffectively, someone—whether Orville Wright, Alpheus Barnes, or another person in the murky management of the last years of the Wright Company is unclear—stopped placing ads urging the purchase of airplanes. Instead, the company used the space it purchased to promote its last remaining successful venture, its flight schools, at the original site, Huffman Prairie Flying Field, near Dayton; in Augusta, Georgia; or at Hempstead Plains, on Long Island. After Orville Wright sold the company, in October 1915, new management resorted to using ads to proclaim that they ran a different sort of Wright Company, one more willing to recognize the usefulness of the rest of the aviation community and to look favorably on the work of other inventors and companies when appropriate. Advertisements for the schools published after October 1915 all emphasized that students would not have to learn to fly with the traditional Wright lever controls. Instead, the new management proclaimed that "Dual Wheel Control [is] Used Exclusively" on the old Wright Bs still employed as trainers. This admission of the unpopularity and obsolescence of the Wright control lever system, which the company had reluctantly begun to abandon

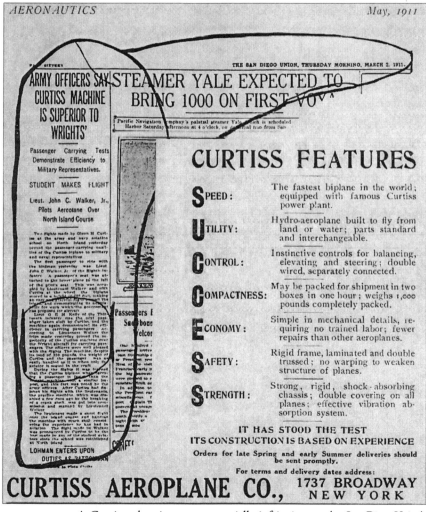

FIGURE 7.3. A Curtiss advertisement, potentially infringing on the *San Diego Union*'s copyright, from the May 1911 issue of *Aeronautics*

as Wright tried to sell the firm, demonstrated a need to advertise the features Wright airplanes did *not* have, not the features they actually employed. A student could learn to fly from the Wright Company's teachers yet not be locked into always flying a Wright Company airplane without additional training, which had not previously been the case. Late Wright advertisements reflected a company with few new products, one whose famous name and patent were its chief commodities. But throughout the company's years in Dayton, its ads generally tried to sell goods or services; they did not try to inculcate readers with particular cultural values.[6]

While the Wright Company shifted from boasting about the achievements of pilots flying its models, in its first years, to emphasizing the 1906 Wright patent and its status (at least to the Wright Company) as the legal foundation of the industry, in its later ads, other builders consistently promoted new designs and technological achievements. Curtiss frequently did so in ads in *Aeronautics* and *Aircraft* between 1911 and 1915, in a much more vigorous manner than did the Wright Company, and with more stylistic panache. The variety of different themes and designs in Curtiss advertisements suggest either the engagement of a professional advertising firm or the work of a dedicated, talented employee in Hammondsport, but their ultimate origin is unclear. Curtiss could be unique in its promotions. The Wright Company never promoted itself by asking readers to send it their ideas on "the future development of the Aeroplane Industry" in return for an autographed photograph of Wilbur or Orville Wright. Curtiss promised to send a signed photo of its namesake to those who responded to its ad in the December 1911 issue of *Aircraft*. Curtiss also tried to capitalize on the fame of its founder outside the aeronautical community by also running the ad in the January 1912 issue of *Popular Mechanics*, a magazine with a monthly circulation of two hundred thousand, a broader audience even than *Town and Country* (though Curtiss also advertised in its "Air Scout" supplements). Subsequent issues of the magazine included ads for the Curtiss flight school at Hammondsport, but the company also continued to advertise its airplanes. In addition to advertising in more widely circulating magazines, Curtiss used newspaper coverage of its airplanes in its promotions. In 1912 it ran pieces in *Aircraft* and *Aeronautics* that included a photographic clip of a *San Diego Union* article where some army officers claimed that Curtiss airplanes were better products than those of the Wright Company, in speed, safety, and overall strength. Curtiss may have violated the *Union's* copyright on the article, but the potential of a relatively minor copyright infringement case was immaterial to a company then fighting the Wright Company tooth and nail in federal court on patent infringement.[7]

Still, the most active aviators read the aviation monthlies in addition to *Popular Mechanics*, and Curtiss made sure to keep its wares before their eyes. Between 1912 and 1914, Curtiss boasted about the features of its new hydroplanes in the ads it placed in the aeronautical press, both to capitalize on the demand for seaplanes and to focus public attention on products that it viewed as less subject to the Wright Company's patent infringement lawsuits. Curtiss advertisements emphasized the ease of use and safety of their models, advising potential customers that airplanes from Hammondsport were "as easy to operate as a motor car" and that in 1913 there was not "a single serious accident" in more than two hundred thousand passenger-miles flown in Curtiss flying boats. In late 1914 aviators were no longer setting records in Wright airplanes, but they were achieving new feats with Curtiss products. That company's ads made sure readers knew about the altitude record of more than seventeen thousand feet set by army captain H. Leroy Müller in its new tractor biplane, the Model J. By 1915 growth and stability were at the forefront as the company gave notice of its new factory in Buffalo, with a photograph of the production floor of "the largest and best equipped aeroplane manufacturing plant in the world." With its corporate stability established, and with the new aviation needs at the start of the First World War, Curtiss advertising turned to promoting the powerful engine, speed, and weight-carrying abilities of its Model R military tractor biplane and the company's line of motors. The range of advertisements, their placement (including the cover of *Aeronautics* between September 1913 and the publication's final issue of 30 July 1915), and the company's willingness to market itself in *Popular Mechanics* to a broader readership than the relative few who subscribed to the trade journals are representative of a company much more active in attempting to massage its public image than was the Wright Company.[8]

Frank Russell may have taken some of the Wright Company's conservative publicity philosophy to Massachusetts when he became general manager at Burgess. The Marblehead company's advertising resembled the Wright Company's more than it did Curtiss's. While it also joined in marketing itself in *Town and Country*'s "Air Scout" section (Starling Burgess was the sort of person—a member of the well-to-do Eastern establishment—who would take the magazine generally), Burgess did not advertise in more general publications such as *Popular Mechanics*, and it did not change its ads as often as did the company from Hammondsport. However, it continually emphasized the qualities of its airplanes and the records pilots set with them. Interestingly, Burgess minimized the significance of its licensing agreement with the Wright Company. In *Aeronautics*, only three ads in the spring of 1911 acknowledged

FIGURE 7.4. One of Curtiss's front-page *Aeronautics* advertisements, from September 1914, featuring Müller's altitude record. One of the issue's original owners was Lt. A. F. Bonnalie, a Californian attached to the British Royal Air Force, who received the Distinguished Service Order in 1918. E. W. Robischon wrote about aviation.

that relationship and proclaimed that the "Wright Company, on account of the excellence of our construction, invited us to become the first manufacturers in America licensed to build aeroplanes under their patents," while small type informed *Town and Country* readers that Burgess was "licensed under the Wright patents." Achievements mattered more than licenses. To promote the firm, Burgess emphasized Harry Atwood's long-distance and endurance flights in a Marblehead-built airplane in advertisements in late 1911 and early 1912, as well as an endurance record for pilot and passenger of almost four and a half hours, set by army pilots Thomas DeWitt Milling and William C. Sherman in the spring of 1913 in a Burgess military tractor biplane. The company frequently announced the military capabilities of its aircraft as the shadow of war-torn Europe. Burgess was proud that the U.S. Navy had acquired its flying boats, which held "the records for best military performances" and "represent[ed] a startling departure in construction"; civilians could also acquire models for sporting purposes. In late 1914 (or early 1915, since *Aeronautics*, which was soon to disappear, did not publish its 30 October 1914 issue until 11 February of the next year), Burgess reported that its Burgess-Dunne military aeroplane, "par excellence the weight and gun-carrying aeroplane of the World," was in the arsenals of the United States, Great Britain, and Russia. Even as Starling Burgess negotiated his company's acquisition by Curtiss Aeroplane, a full-page advertisement in *Aerial Age Weekly* trumpeted its role as a supplier "to the United States Army and Navy and to the British Admiralty" of "military planes for national defense." Burgess workers were not solely building military airplanes, though; the same notice reminded customers that the craftsmen in Marblehead still built "sea planes for sport." But these purchases were not enough to preserve Burgess's independence. Profits were not what Starling Burgess desired, and Glenn Curtiss needed more production space. In February 1916, Curtiss acquired the Burgess Company; as a division of Curtiss, Burgess continued to build (and advertise) airplanes until 1918.[9]

Did the Wright ads work? They did not work particularly well, given how few airplanes the Wright Company built overall. But advertising did not drain the company's treasury. It was not even considered a significant expense worth recording on its own in the company's ledger. The same volume does contain entries for accounts receivable and for sales, and these pages depict a marked decline in corporate fortunes (and in item-level detail) in the years after 1912, as the company's products aged and as consumer demand for seaplanes and tractor airplanes—products where other builders dominated—grew. In 1910 and 1911 the company recorded the name of every customer, whether for spare parts,

FIGURE 7.5. An early Burgess advertisement, from the March 1911 issue of *Aeronautics*, boasting of the company's license agreement with the Wright Company

engines, or airframes. The army's purchase of Model Bs, the licensing arrangement with the Burgess Company, and Cal Rodgers's acquisition of his *Vin Fiz* EX all appear in its pages. And then Frank Russell left Dayton. The company's skill in bookkeeping declined significantly after the former Automatic Hook and Eye president took his business background east. After 1911 the company made only unnamed dollars-and-cents entries, identifying each in the ledger with a number (one number could be attached to multiple entries, suggesting that it identified a particular customer, though no key survives). From over $100,000 in sales in 1910 and 1911 (years combined in the ledger), and $120,000 in 1912, sales income fell to just under $26,000 in 1913 before recovering to slightly over $26,000 in 1914 and $40,000 in a shortened 1915 (due to Wright's sale of the company that October). Over time, income from other sources also declined. Royalty earnings peaked at $35,616 in 1912 before falling to $2,163 in 1914 and, according to the ledger, a rather curious zero in 1915. License fees from exhibition aviators, which the company found difficult to collect (which was why Alpheus Barnes pursued Cal Rodgers into New York during the first days of the *Vin Fiz* trek), peaked at $8,850 in 1911. That same year, the company collected $7,000 in fees from manufacturers, principally Burgess. Exhibition earnings accrued only in 1910 and 1911, while pilot lessons brought in several thousand dollars each year between 1910 and 1915. In 1911 the company also earned interest on New York City revenue warrants. The Wright Company disbursed four dividends to its stockholders, one in 1910, two in 1911, and one in 1912, but none in 1913 (and with Orville Wright taking control in 1914, the need for dividends ceased). The company often found it difficult to achieve enough financial stability to meet the payroll. The Wrights personally advanced the firm several hundred dollars on more than thirty different occasions between 1910 and 1916 to pay workers. They even made unspecified loans of $1,300 and $1,000 in 1911 and 1913, respectively, to provide the company's accounts some cushion. Toward 1914, with the company's product-related revenues anemic, the directors in New York started to complain about the fate of their investments.[10]

The directors had reason to be worried if they read of company affairs in *Aeronautics* or the New York press, or stopped by the New York office, for other airplane makers were expanding their facilities and selling more planes. In Hammondsport, business was booming. While in 1914 Curtiss's airplane was still not as profitable as his separate engine-making and motorcycle businesses, war in Europe created several overseas markets for his company. Curtiss, based in the neutral United States, did not play favorites. It exported airplanes to Russia and Japan, of the Entente Powers, and Germany, of the

Central Powers. Curtiss also continued to sell airplanes to the U.S. military, and especially the navy. By October 1915, when Orville Wright sold the Wright Company, Curtiss was the largest airplane maker in the United States. European demand for its planes grew sufficiently that its Hammondsport factory was too small to produce the number of aircraft demanded in Europe, or to provide space to the number of workers necessary to build them even though it had rarely operated at full capacity. It was also in a small town with limited shipping options and railway connections. Therefore, Curtiss moved most of his operations from Hammondsport to Buffalo, ninety miles northwest of his hometown, in late 1914 and 1915, where he opened a large, new factory, as well as to Toronto, on the north side of Lake Ontario, where it would be easier to directly supply British forces from what was then more directly part of the British Empire. Only the war kept facilities open in Steuben County; the Hammondsport factory closed permanently after the war ended. In the autumn of 1915 an investment banking firm, William Morris Imbrie and Company, purchased controlling shares of the three companies for $5 million in cash; Curtiss himself remained in day-to-day control while also receiving $3.5 million in preferred stock, much more money than Orville Wright would be paid for the Dayton firm. In 1915, Curtiss had a bright future.[11]

Meanwhile, business in Marblehead was less promising than in Hammondsport. Starling Burgess had reorganized his company (now simply renamed the Burgess Company, though Greely Curtis remained its treasurer) after its licensing agreement with the Wright Company ended. The company embarked on a new licensing agreement with British aviator John Dunne, building versions of his airplanes that gained some popularity in the civilian and military markets. Burgess airplanes even appealed to Orville Wright's investors. Keenly interested in new developments in aviation, Wright Company director Robert Collier, who would retain his company stock right until the 1915 sale, purchased a custom-designed seaplane from Marblehead. However, even the commencement of the Great War did not generate enough sales for the company to be consistently profitable, and in February 1916 Burgess sold it to Glenn Curtiss for $900,000. Curtiss, whose company needed the extra production capacity, made it the Burgess division of the Curtiss Aeroplane and Motor Company. Starling Burgess left Massachusetts for a lieutenant commander's commission in the U.S. Navy Bureau of Construction and Repair in Washington in 1917; the company's workers continued to build airplanes until a November 1918 fire destroyed one of the Marblehead factories. Curtiss did not reopen the factory, and Starling Burgess returned home and to

FIGURE 7.6. A Burgess advertisement from the October 1914 issue of *Aeronautics*, placed after the dissolution of the company's license agreement with the Wright Company—and after Frank Russell had moved to Marblehead

yacht building after the war. Meanwhile, the war benefited a California company that later had a fleeting connection with the postsale Wright Company. The business of aviator Glenn Martin, organized in the summer of 1912, went from a small operation employing fourteen men in early 1913 to one producing dozens of airplanes that attracted inquiries from the militaries of the United States, the Netherlands, and Great Britain in early 1915. But none of these developments—in New York, Massachusetts, or California—affected Orville Wright's business approach. In 1914 he fought off efforts from the Wright Company's board of directors to strongly enforce the company's patent lawsuit victory, bought back most of their stock, and began the process of leaving behind the life of a corporate president. Both brothers had always been more interested in the workshop than the office, and they found occupying the company's presidency to be a difficult task. Though they initially attempted to delegate daily operations, they found that delegation was not to their liking. The men they and the board of directors hired to manage the company's day-to-day operations—Frank H. Russell in 1910 and 1911 and Grover Loening in 1913 and 1914—found that their attempts to manage the company with Wilbur and Orville Wright as president (and, for Russell, as vice president) met with significant resistance. Relations with Alpheus Barnes, who ran the New York headquarters, were little better.[12]

# Managing the Wrights' Company

Early Wright Company advertisements boasted that Wilbur and Orville Wright personally supervised the designing and building of "everything that enters into the construction of our machines" at the factory. For once, advertisements did not lie. The brothers knew that the Wright Company was a different sort of business than their previous printing and bicycle ventures, and maintaining such close, personal oversight was not their stated intent during the company's first days. In December 1909 they wrote to accountant and future Wright School student and aviation designer Albert Merrill, "We have a considerable stock in the company and are serving in official capacities, though we will not have the care of the general business," suggesting that they recognized at some level their lack of experience in running the sort of company they hoped would be a major firm of national scope. Professional management, it seemed, would be required, and the company proceeded to hire Frank H. Russell (1910–11) and Grover C. Loening (1913–14) to fill that role. But Russell and Loening would find that the Wrights' statement to Merrill did not reflect reality. Instead, Wilbur and Orville Wright, the only Wright Company executives who lived in Dayton, became company executives unwilling to let others manage their business. Russell and Loening also found the New York–based corporate secretary and treasurer, Alpheus F. Barnes—who had his own management conflicts with the Wrights—a difficult man for whom to work and attempt to craft a profitable enterprise.[1]

Difficulties between Russell, the Wright Company's general manager from January 1910 until October 1911, arose early in his tenure in Dayton. Russell's more privileged upbringing and education, his more extroverted personality, and his previous corporate experience set him apart from the Wrights. He was born in Mansfield, Ohio, but raised principally in the Northeast, where his Congregationalist clergyman father worked as secretary for the Evangelical Alliance for the United States of America and led congregations in New York and Connecticut. His maternal grandfather, Russell A. Alger, a Republican, served as governor of Michigan, William McKinley's first secretary of war, and U.S. senator; Wright Company director Russell Alger, Jr. (Russell A. Alger's son) was his uncle (though only five years his senior). A 1900 Yale College graduate who gained business experience working for his uncle in Canada, Russell arrived in Dayton from the presidency of the Automatic Hook and Eye Company, a zipper and fastener manufacturer in Hoboken, New Jersey. Automatic Hook and Eye was faring poorly in the aftermath of the Panic of 1907, and Russell, who was owed over $8,000 in back pay, was eager to find a new opportunity. He quickly took the managerial reins in Dayton when they were offered to him. He became the only Dayton-based company official with previous experience in running a corporation to ever have a significant role in the Wright Company's daily operations. When hired, he understood that the Wright Company wanted him to "have general oversight of details of construction sales contracts etc." Indeed, the company had told him as much in his appointment letter from the brothers, which placed him under the "general direction" of the company president but expected him shortly to "assume full charge of the management of the company's business." But Russell quickly learned that he was working for two headstrong brothers, men who were unable to put their words to him into action.[2]

Letters and contracts aside, Russell was never given full charge of company management. The corporate chain of command bypassed him, leaving him continually frustrated. During his time working in Dayton he found the Wrights meddlesome and critical supervisors who provided little constructive guidance and interfered in his efforts to create a profitable business, while the brothers found him to be incompetent. While remembered by machinist Tom Russell as someone liked by the workers on the factory floor, his workplace personality jarred with the introverted Wrights. As Russell left the company, in 1911, his uncle wrote rather cryptically that he (Alger) could "quite easily see where Frank Russell's personality has to a certain extent worked against him in his work" and that he agreed with Wilbur Wright's view that Russell "disturb[ed] the general organization." Russell, however, believed that Wright disturbed *his* organization. Earlier that year, Russell wrote in his diary that Wilbur Wright claimed that he (Russell) had not shown "enough 'genius' to warrant his giving me the business," a comment that prompted him to tell Wright that "constant criticism & suspicion don't promote 'genius.'" But Russell managed to leave a large portion of this criticism at the office. Marietta Russell, Frank's wife, socialized frequently with Katharine Wright, and Bishop Milton Wright's diary contains more than forty entries of the Russells dining

FIGURE 8.1. Frank H. Russell (*left*) with Major General Mason M. Patrick, Chief of the U.S. Army Air Service, 1922. *Courtesy of the Manufacturers Aircraft Association, Inc. Papers, American Heritage Center, University of Wyoming*

with the Wrights, or of the Wrights dining at the rented Dayton home of the Russells. Amazingly, after resigning, Frank Russell even lodged with the Wrights at their cramped 7 Hawthorn Street home after his family's furniture had been packed for shipment to Massachusetts, and Katharine Wright traveled with his wife and children to Mannsville, New York, his wife's hometown.[3]

The Wrights' inexperience in creating and managing an effective organizational structure also affected Russell's relationship with Roy Knabenshue. While the workers on the factory floor may have worked easily enough with the former zipper manufacturer, Russell quickly came into conflict with Knabenshue, a Wright favorite. Russell and Knabenshue clashed over the delegation of management authority delegated to both men from the company's board. No one thought to draw or enforce an organizational chart. While Wilbur Wright reemphasized the company's appointment letter to Russell in July 1910, confirming to him that as general manager he was responsible for all the Wright Company's departments, Wright told Russell a month later that "Knabenshue's agreement gave him charge of [the] Ex[hibition] Dept. and that I had interfered much to the gentlemen's discomfort." Wright and his board of directors were unable to clearly delineate the specific oversight responsibilities for Russell and for Knabenshue upon each man's hiring. In conversing with Russell in August, Wright also seems to have forgotten his letter of hire to Roy Knabenshue the previous March in which he told the Toledo aviator that Knabenshue had "charge of all of the details" of the exhibition department and that he was to confer with "us and the Board of Directors with reference to the policy to be pursued" and not with general manager Russell, who was not consulted concerning the hiring of an exhibition department manager. Russell was not pleased with what he saw as a curtailment of his authority and a vote of no confidence by a company president acting without board approval, writing in his diary that he "asked [Wright] in return for relinquishing all control of that Department that he let me know as soon as possible whether my services were such" that he would not be required after 1910. Nor did Knabenshue care for Russell, who he felt unduly interfered in exhibition matters; Russell signed several contracts for exhibitions (usually one of Knabenshue's duties; it is unclear whether this reflected Russell's specific intention to assert authority, or whether he simply signed in place of an absent Knabenshue) and supervised the company appearance in Aurora, Illinois, in June 1910. When the time came for Knabenshue to renew his contract for the 1911 exhibition season, in March of that year, he insisted that the company's general manager have no role in exhibition affairs. With the exhibition business much more profitable than sales of individual airplanes, Orville Wright agreed with Knabenshue's request and placed him under the direct supervision of the president and vice president. Russell, his shrunken status in the front office evident, now truly had no authority over the aviators working under direction from the thirteenth floor of the United Brethren Building, in downtown Dayton; he served merely as an alternate if Knabenshue was unavailable. The brothers allowed him to travel to Springfield, Illinois, and sign an exhibition contract in April 1911, when Knabenshue was ill with quinsy, a form of tonsillitis. Knabenshue soon recovered his health, enough to try to escort the visiting Glenn Curtiss, who was passing through Dayton, on a tour of the Wright Company factory. Joined by Russell and Orville Wright, Curtiss had the gall to ask the Wright Company to pay Curtiss for aviators who were flying Curtiss airplanes without paying royalties. "It's an outrage," Curtiss told his competitors. Russell, fortunately, retained enough authority, according to Katharine Wright, that he ordered the plant "watchman to let absolutely no one in."[4]

As his power over the company's departments eroded, criticism of Russell's managerial style and his desire to be general manager in more than name grew, and that criticism came especially from the Wrights. Existing papers provide little insight into what about Russell's managerial approach irked the Wrights. Having no need to employ a manager in their printing or bicycle sales and repair businesses, and never having worked as employees of someone else or having previously spent significant time within a corporate managerial structure, the Wrights may have had different, uncommunicated expectations of the general manager's ultimate role when they hired Russell than he did after having been president of Automatic Hook and Eye (Russell, no doubt, viewed the role of the company president differently than did the Wrights). Though in late 1909 they may have agreed in principle with the idea of removing themselves from daily company operations, in practice the Wrights were not comfortable with letting someone else have a significant measure of control over the future of a technology on which they had spent more than a decade of their lives. Their criticism of Russell could be caustic. Indeed, by late May 1911, the Wrights accused Russell of insubordination. Writing to Wilbur, who was then in Paris attending to the brothers' European affairs, Orville sarcastically noted that the "'organization' of our shop is being gradually perfected by the business manager," who thought "he was to have complete control of the business and was to be under no one whatever excepting you!" Russell, it seems, still wanted to assert his role in the company as an official with duties as "those of general manager under the general direction of the president," as stated in his letter of appointment. Russell maintained a residual hope that he might somehow do something that would make the Wrights' early 1910 statement that they "hoped that in a short time you will be able to assume full charge of the management of the company's business" more than a forgotten platitude. At his hiring, Russell was led to believe that he would be the principal person in Dayton with day-to-day managerial responsibility for the company, with the company president and board providing big-picture guidance but steering clear of workaday matters. However, by the summer of 1911, the Wrights considered Russell's empowerment a relic of their company's first months. They were no longer willing to give him full charge of the company's business. Instead, the brothers saw him as a glorified foreman, responsible only for the internal operations of the Dayton factory and directly answerable to both brothers, not just the company president.[5]

Did the Wrights have cause to accuse Russell of insubordination? The paper trail of his relationship with the brothers is not extensive, since frequent interaction in Dayton made letter writing unnecessary for all but the most formal events, for issues when a party wanted to create a written record, or if a party happened to be away from western Ohio. It is clear, though, that by 1911 the brothers and Russell had frosty relations. Russell's writing shows a frustrated, defiant general manager, one who was trying to use the distance between the New York directors and Dayton to his advantage as his rapport with the Wrights deteriorated. An issue with the company's production capacity demonstrates this situation. During the first half of 1911, the Wright Company encountered difficulties fulfilling Burgess Company orders for motors for licensed Burgess-Wright airplanes, while building sufficient motors for its own production needs. The company ordered the construction of the second factory building to help alleviate these production problems, but it did not open until the late fall, after Russell left Dayton. Throughout the heat of June and early July, Orville Wright discussed the matter several times with Russell, telling him that the company needed to acquire additional machinery to alleviate the order backlog. Meanwhile, director Russell Alger had ordered a Model B, and he made a visit to the factory to see the works and to check on the progress of his machine—and to try to mediate between the Wrights and his nephew. Alger met with Wright for two hours, finding him to have set opinions on the status of the company. He advised his nephew that he was unsure whether his visit was of any use, but that Russell should wait and discuss matters with Wilbur Wright upon his return from Europe. He did not advise Russell to take the action he took the next day. Responding formally to his criticism, Russell sent Orville a cold Dear Sir letter to the 1127 West Third Street office. He wrote that the acquisition of additional motor-making machinery would not help the company fulfill the orders; indeed, he had recently "added a spare lathe, drill press, [and] running[-]in attachment for motors" that would allow the production of seven or eight motors each month. The contract with Burgess called for three motors to be sent to Marblehead, leaving four or five (according to Russell's estimate) for use in Dayton airplanes, which should have sufficed for the number of airplanes the factory workers could build for the Wright Company itself—and Burgess did not make any complaint to the Wright Company about delayed orders. Rather, after discussing the matter "again very fully" with longtime Wright mechanic Charles Taylor, Russell claimed that any Wright Company manufacturing problems resulted from an "almost impossibility of securing trained men for our work, especially assemblers." Russell disputed that Marblehead would even have reason to complain, claiming that "we have been able to keep up with" the Burgess orders. Russell endangered his position with this letter; he did not know that Wilbur had informed Orville in May from Europe that "if he [Russell] gets smart on your hands 'fire' him." He closed his letter by bluntly acknowledging Wright's supervisory role. "If the above is not in accordance with your own judgment, I should appreciate very highly any suggestions concerning any equipment which you might wish to have installed, at once," he wrote, "and your judgment will be followed in this matter as in other matters in the past." Russell, to whom Wright must have expressed his wishes many times over the previous months, had come to require more formal direction than a mere workplace conversation. He wanted written documentation.[6]

Meanwhile, Russell damaged his standing with the brothers by telling the New York office and vice president Andrew Freedman another version of the Burgess motor supply story. After sending his letter to Wright, Russell met with the vice president to confer about suing aviator Earle Ovington, who flew Blériots, for patent infringement (Russell's diary does not reveal whether he and Wright discussed his letter). He also discussed matters with navy captain Washington Chambers, who arrived in Dayton to inspect a Model B ordered by the service. He then traveled to Mannsville, New York, for a brief holiday before proceeding to New York City, where he met with Freedman on 26 July. Freedman was not happy. He questioned the company's frequency of production, its anemic sales figures, and its (in his view) insufficient resort to the courts to protect its patent. In a later meeting that included De Lancey Nicoll and Alpheus Barnes, Freedman told Russell that he did not believe that the latter was keeping them and their colleagues on the board of directors informed of company operations, especially its legal matters. Russell discussed his perception of Dayton operations with the men and followed his discussion with a letter to Freedman on 28 July, providing his oral advice in a more formal manner. Russell told the New York–based vice president that the speed of airplane production was limited by the company's capacity to make motors, and not, as he had told Orville Wright two weeks previously, by staffing levels. Workers could build a maximum of six motors and nine airframes a month in the existing facility, he wrote. Russell now believed that supplying three motors per month to Burgess "has handicapped our production of complete machines" and that "further equipment" was required to increase production. Sales, according to Russell, were lackluster since they were not handled in a systematic manner. He urged the board to authorize him to hire a "competent salesman" to handle sales, run a publicity campaign in both trade and popular magazines, and open a sales office in New York City, where agents could "talk with prospective purchasers." Reducing the price of exhibition royalty licenses from the existing level of $100 a day to either a percentage of a pilot's gross receipts or a flat $25 or $50, Russell believed, would make aviators who bothered to pay royalties more competitive with wildcat aviators who paid no royalties in scheduling exhibitions (and perhaps they would fly Wright Company airplanes). Lower fees would also encourage wildcatters to take out licenses and pay royalties and thereby increase company revenues. Russell then made a curious statement. The company needed to appoint a staff attorney to vigilantly protect the Wright patent, he wrote, and to sue for infringement "all the important exhibitors and professional flyers" and the various builders (especially the Farman and Moisant firms) attempting to open factories. As Wright Company manager, Russell believed that additional lawsuits would "aid our sales and justify our license fees ... more than the resulting criticism will hurt them." Was Russell so marginalized or so oblivious that he was unaware of Wilbur Wright's frequent trips east or of Harry Toulmin's patent law work, a short distance away in Springfield? Did Russell think that De Lancey Nicoll spent too little of his time on Wright Company work? Experimentation, or the lack thereof, also troubled Russell. He advised Freedman that the company also needed to conduct more experimental work to develop new models that responded both to government directives to install pontoons on airplanes and to the Curtiss Company's successes in water-based aviation, which was "the strongest inducement they are making for the sale of their product." The ability to experiment was hampered by the firm's lack of a staff engineer. Russell, perhaps naively, wanted Orville and Wilbur Wright to allow others to improve the designs of the company's airplanes. He wanted to hire an engineer "to work under the advice and direction of Mr. [Wilbur] Wright and relieve him of the details in connection with improvements." Given their close control of the design of company products, it is unlikely that either Mr. Wright—neither of whom (and Russell did not specify which brother he meant) was a certified engineer—would have delegated much work to this person, just as they turned away from delegating much control of the company to Russell himself. Russell closed his letter requesting that the board establish "a more clearly defined policy" to ensure that the company's "entire efforts may be centralized on the production and sale of a thoroughly successful aeroplane." The Model B then in production was a successful airplane, but Russell worried that the company had little under development that would interest its prospective customers or compete well with new Curtiss or European models.[7]

Russell would not be happy with the meeting's result of stasis. Writing about the general manager's visit to Orville Wright on 28 July, a Friday, the Freedman told the inventor that the meeting produced a consensus between him and De Lancey Nicoll concerning the best direction of the company's affairs. The Wrights, wrote the former owner of the New York Giants, "should be protected in your patents, and in your rights to operate your business." Freedman and Nicoll both thought that Wright "should protect [his] interests as well as the Company's" if he found that patent infringements were causing him serious interference and "where the business is suffering the consequences." Concerned, though, about what he and Nicoll had heard from Russell, and worried that Wilbur would tarry in Europe, Freedman requested that Orville hurry to New York the next Wednesday, 2 August, to meet with him and with Nicoll before Freedman left the city for a long vacation in New Hampshire. Orville undertook the jarring train trip, which he generally tried to avoid making so as to not exacerbate his sciatica, and met with Russell toward noon on Friday, 4 August, to allay his concerns. Freedman need not have worried about Wilbur tarrying in Europe, as the elder brother arrived in New York aboard RMS *Oceanic* on 9 August. Satisfied that the brothers were in control in Dayton, Freedman turned his attention to his many other business and political affairs in New York, especially his work with Interborough Rapid Transit president Theodore Shonts on the development of the city's subway system. However warranted, Russell's opinion was ignored.[8]

Had Russell's ideas gained more traction, they may have helped the Wright Company become a more competitive airplane builder. A "competent" engineer might have provided the company additional intellectual resources and could have collaborated with the Wrights in updating their airplane design—though whether the Wrights would have been willing to allow such a collaboration before the validation of the 1906 patent in federal court, or without insisting on idiosyncratic changes, is debatable, as Grover Loening later discovered. And Dayton was not lacking in engineering talent. Orville Wright would be one of the founding members (and third president) of the Engineers Club of Dayton, in 1914. But aeronautical engineering was an infant subfield, and the Wright Company did eventually hire Loening, the first person in the United States with an aeronautical engineering degree, in 1913. Russell would have found much the same situation existing in Hammondsport. At Curtiss, blueprints—when they were used at all—were "rudimentary," and workers "designed as we went along and there was a lot of guesswork." Back in Dayton, contractors completed construction of a second factory building in the autumn of 1911, increasing the amount of available production space for men and machinery, but one of the company's largest customers, its exhibition department, simultaneously disbanded, reducing demand and the need for more men or machinery. Sales of the Model C in 1912 were not sufficient to replace this internal demand, and new machinery would have been underemployed without the development of a new model that was in high demand. Wilbur Wright later noted in a letter to Orville that Alpheus Barnes had "put up a sign giving the location of our N.Y. salesroom." This might have made it easier for potential customers to procure an airplane. Whether it did, though, is not clear—nor is its location, its hours of operation, the number of sales transacted, or whether the "salesroom" was just another corner of Alpheus Barnes's office. Company advertisements and city directories for New York City are mum on the matter. Wilbur Wright acknowledged that, in the crowded exhibition arena that the Wright Company soon left, purses for honest exhibition aviators were falling and pilots were finding it difficult to pay the royalties due to the Wright Company, making collections "slow." But slow was acceptable. Unlike Russell, Wright was a direct beneficiary of the royalties, and he did not suggest lowering them.[9]

Frank Russell did not wait for the company president's response. He felt strongly enough about the implications of royalty collections on the company's affairs that he sent a telegram directly to Orville Wright in Dayton right after his meeting with Freedman. Russell, visiting his old Automatic Hook and Eye colleagues in Hoboken, claimed that promoter William ("Jim") Gabriel, who would manage former Wright exhibition pilot Leonard

Bonney, was refusing to accept the plane he had ordered from Dayton that August without a "withdrawal" of the royalty fee assessed to exhibition aviators. Russell believed that the fee handicapped honest pilots; wildcat pilots who refused to pay the royalty could charge lower appearance fees. The Wright Company, Russell advised its vice president, needed to waive the royalties charged to pilots until after the courts adjudicated the Wrights' patent infringement claims, or "we are not going to be able to sell our product" to independent aviators. Wright would not budge. He responded to Russell several days later, saying he did not "think it advisable" to change the royalty policy. Gabriel's threat turned out to be bluster, as he paid $5,000 for his Model B on 18 August 1911 (though whether he ever paid royalties is unclear). The Wrights had decided that these policy issues were their responsibility, not the responsibility of their general manager.[10]

Unable to implement his proposals for corporate growth, the Frank Russell who returned to Dayton at the beginning of August was a diminished general manager. He began to evaluate his future with the company and in aviation. For a few weeks he busied himself with routine correspondence with Alpheus Barnes, discussing company accounts, expenses, and royalty issues for an exhibition appearance at the Chicago International Air Meet, and sales contracts. He traveled to Chicago to meet with the managers of the air meet, who were concerned that the Wright Company would sue to keep the event from opening (it did not), and spent a few days with his family in Mannsville. He also entertained new employment opportunities, for the Wrights had decided to part company with him and run the business themselves, without a general manager. Meeting with Wilbur Wright on 6 September, Russell agreed to resign, effective 31 October. He did not have to worry about extended unemployment. Within two weeks, he had a job offer from New England, from a Burgess Company that was familiar with his work in Dayton through their licensing agreement with the Wright Company and that wanted to use his experiences with that firm in improving its own line of airplanes. Expressing to the brothers hopes of "continuing our friendship which I value most highly" in his new job with a company "whose aims," he thought, were "the same and whose interests are so nearly identical" with those of the Wright Company, Russell became the manager of the Burgess Company and Curtis on 1 November. For his work in Dayton, the factory's workers gave him a watch; in return, he took twenty-four of them to a performance at the Lyric vaudeville house on East Fifth Street on his last night in town. He then boarded a train to Mannsville, where he would relax with his family before heading further east. Russell did not travel to New York alone, however. Always fond of the Russell family, Katharine Wright accompanied him on the overnight train, spending three full days with the reunited clan before beginning her return trip to Dayton, with stops in Geneva, New York, and at Oberlin College. While in Mannsville, she telegrammed Bishop Wright that everyone was well and that "all the Russell [sic] send love to Grandpa Wright." While Milton and Katharine were sad to see the Russells leave, Wilbur and Orville were less bereft, and received a sympathetic letter from Russell's uncle, company director Russell Alger. He wrote to Wilbur that while Frank Russell's "personality has to a certain extent worked against him in his work," he was "a splendid young man and with his energy and undoubted intelligence should do well." To his nephew, Russell Alger had been consoling, telling him during a September meeting in Detroit "how awfully bad business was" and how he was "not hopeful for the aeroplane business." Frank Russell, though, was hopeful and was happy to leave Dayton.[11]

In Marblehead, Russell became an important leader of the Burgess Company in the years before its acquisition by Glenn Curtiss. He gained partial ownership of the company, receiving $4,800 in salary ($800 more than his final Dayton salary) and five thousand shares of stock. The Burgess Company had different needs than its Dayton competitor, but Russell ensured that readers of the aeronautical press took notice of its products, placing a few promotional articles in *Aircraft* in 1913 and 1914. Russell remained active in transportation for the rest of his working life. He gained additional influence in the industry as a vice president of Curtiss, from 1920 to 1931, participating in the development of the NC-4 flying boat, and as an officer of the Manufacturers' Aircraft Association, the industry body created in 1917 that managed the aviation patent pool and disbursed royalties to its members. From 1931 to his 1936 retirement, he was a vice president and director at the Edward G. Budd Manufacturing Company, which built stainless-steel railway cars and automobile bodies. He died at his Bucks County, Pennsylvania, home in 1947. With Russell's departure, the Wright Company lost the only Dayton-based manager it ever employed with a background not in engineering or in craft work but in business management. His departure left the Wright Company managerially adrift as it began its decline as a significant U.S. aircraft manufacturer—and left the Wrights in complete control of daily activities at the Dayton factory.[12]

Shortly after Russell resigned, Wilbur Wright wrote to his younger brother from New York City, where he was supervising the progress of the patent lawsuit against Curtiss, asking Orville to "take charge of the factory so far as determining its policy and management." Shop superintendent Arthur Gaible assumed some of Russell's local duties on a long-term interim basis from late 1911 to the summer of 1913. The Wrights' personal financial records also suggest that local businessman Alfred G. Feight received payments for an unspecified "superintendence" for a few months in 1912. The Wrights were comfortable in their relationship with the directors in New York and with the plant's existing operations. No one in the company hierarchy saw a need to hurry and bring on a new general manager—board meetings ignored the issue entirely. Russell Alger remained silent on his nephew's treatment by the Wright Company and his move to one of the company's principal competitors. The general manager position continued to lapse. Indeed, the company never hired anyone to fill Russell's particular job title. The New York directors, not yet overly concerned about the company's future and their investments, acquiesced. The company's slide to irrelevancy was just beginning.[13]

The company was midway along that slide before Orville Wright decided to make a change in the summer of 1913. His change, though, was not a re-engagement with someone with significant corporate experience. With a year of experience as company president since Wilbur's death, Orville decided to hire one of the country's best-educated young engineers into an undefined but part-managerial and part-"junior Wright" role. In the summer of 1912, Grover Cleveland Loening, who earned a bachelor's degree from Columbia University in 1908 and was subsequently awarded the first master's degree in aeronautical engineering given in the United States, in 1910, wrote to Orville Wright in search of a position in which he could apply his design experience, much of which concerned hydroplanes. Loening had met Wilbur Wright at the Hudson-Fulton Celebration in 1909. There he helped clean away oil dripping from the airplane Wright flew, an experience he later described as "the foundation stone of my career as an aeronautical engineer." Like Frank Russell, Loening came from a well-off family. His broker father, Albert, served as U.S. consul general in Bremen, Germany, where his son was born during President Grover Cleveland's first term (hence his son's name), and the Loening family had enough wealth to employ two Austrian immigrants as servants in their Manhattan home. After leaving Morningside Heights, Loening wanted to apply his book learning to actual airplane construction and to work with a recognized mentor. He was less interested in finding a highly paid job, proposing instead to Wilbur an arrangement that in a later century might have been considered an internship. Due to his "youth and the position of my family,"

Loening advised Wright that he "considered the matter of salary quite subordinate" to the opportunity to work with someone of his status. Loening's 1912 letter, spurred by Wilbur Wright's death, was poorly timed. It reached Dayton mere days after Wright's burial in Woodland Cemetery, and a bereft Orville Wright waited a month to reply. When he did, he told Loening that the Wright Company had no position for him but that it would keep him in mind for future openings, and in 1912 Loening went to work for the Queen Aeroplane Company in New York City as its chief engineer for $300 a month. Owned by Chicago stockbroker Willis McCornick, Queen produced copies of Blériot airplanes. Loening recalled McCornick as being a playboy, but also a visionary who predicted the eventual adoption of streamlined metal airplanes with enclosed cockpits in a time of exposed pilots flying fabric-and-wood planes, and as a company owner who would financially support the development of such technologies.[14]

Indeed, Loening believed that McCornick spent too freely. He wrote that he "saw money squandered hopelessly, impractically, in the most offhand manner" and gained the enmity of "a motley assortment of promoters, cranks, and 'inventors,'" including the aviator James V. Martin, who went on to unsuccessfully sue much of the aviation industry for uniting to "drive [his Martin Aeroplane Factory] out of business" in the 1920s and 1930s. Queen Aeroplane, though, allowed Loening the opportunity to experiment with hydroplane designs. In June 1913, Loening took his latest creation to a beach near Hoboken to test. While he had high hopes for his airplane, an afternoon thunderstorm destroyed the craft, its winds turning his hopes into "a shapeless mass."

FIGURE 8.2. Grover Loening (*left*) and Orville Wright, ca. 1913. *Courtesy of Ralph S. Cooper/CHIRP*

Emotionally drained by the airplane's destruction, Loening soon found cause for joy. The rapid development of the market for hydroplanes over the past year had changed Wright's position on Loening's potential value to the company. With a year of paid experience at Queen's Fort George factory under his belt, Loening again wrote to Wright, telling him of his desire to apply his ideas on airplane design commercially in Dayton and his wish of having "a small part in the great work of your brother and yourself." He did not ask to be given a managerial role. Wright, in New York for a Wright Company board meeting, met Loening in his home city, and hired him in July 1913. Loening, not wanting to rely on family money (and not offered travel support by the Wright Company), pawned his watch to afford the railroad fare to his new job in Dayton.[15]

Loening's hiring demonstrated the low priority Orville Wright assigned to the management of the company's business affairs by a Dayton-based employee, a portfolio he personally reserved (with Alpheus Barnes handling daily affairs in New York). While Wilbur Wright was principally responsible for hiring businessman Frank Russell to be general manager, Orville hired an aeronautical engineer three years removed from graduate school to lead the factory. Loening joined the Wright Company in a position with "duties . . . such as shall be assigned by the President," becoming Orville's principal aide. While Loening had no official title (nor was he added to the company's letterhead), the aviation press generally referred to him as the company's manager. "Barnes did all the bookkeeping work as well as advertising and negotiating of contracts and left the management of the plant pretty much up to me," Loening recalled. That management power was much more limited and supervised than that initially promised to Russell. Loening came to Ohio with extensive aviation connections already formed. Army aviators Hap Arnold and Roy Kirtland congratulated Loening upon his arrival in Dayton, and even Glenn Curtiss told Loening that he was "probably located satisfactorily" in his new job. However, he had almost no experience in operating a corporation. As chief engineer of Queen he had spent most of his time at the drawing table or in the field with his designs and prototypes. He had little involvement in the company's front office. Still, Loening quickly learned that his engineering skills could not gain attention if the company's products were not being purchased. He became sympathetic with Frank Russell's 1911 unrealized idea that the Wright Company should expand its marketing to remain relevant in the industry, and immediately after assuming his new position, he attempted to rejuvenate company public relations. Loening quickly contacted the trade journals and offered to send them articles about new Wright Company products. The aviation press was pleased to hear from Dayton; Barnes may have handled advertising, but he did little concerning publicity. *Aircraft* editor Alfred Lawson thanked him for his efforts to strengthen the Wright Company's "woefully weak" marketing and suggested that Loening send him "important news matter and data and designs of the Wright machines" each month. While Loening did not write that quickly, several articles under his byline appeared in *Aircraft* and *Aeronautics* during his year in Dayton. In January 1914 the Wright Company board of directors, also dissatisfied with Barnes's work (or lack of work) decided to supplement Loening's articles with the company's first recorded advertising budget: $3,000, under the direct control of President Wright. Whether Orville actually used the money is unclear. The company's print advertisements changed little in frequency, size, or style, and the company ledger is silent on the money's existence and use. Neither Loening nor Alpheus Barnes received authorization to draw on the funds.[16]

The board of directors knew the source of the company's executive power when it offered Wright control of that $3,000. By the time he hired Loening, in 1913, Wright had abandoned any idea of being a hands-off company officer. Loening remembered Wright as being heavily involved in the day-to-day work of the company, even Loening's technical work, where he could easily signal his preferences. Wright, Loening recalled, "directed all of the design work in the shop, even to small metal fittings, and many a time I had designed some detail and made a fine drawing of it, only to find that meanwhile Orville had gone into the shop and . . . he would not only have designed the part, but had it made right there." Such close supervision left Loening, like Frank Russell, with little ability to set an independent course in managing company affairs or in designing airplanes. But the expectations of 1913 were not the expectations of 1910. The company itself was a smaller organization during Loening's tenure than during Russell's. With the exhibition department long disbanded—its office space in the United Brethren Building now occupied by the local representative of Westinghouse Electric—Loening could not have any role in its operations, nor could he develop a professional relationship with Roy Knabenshue, who was flying independently in California. He could, though, try to rebuild the company's relationship with the U.S. military in the wake of the problems with the Model C. He took over Orville Wright's correspondence with army and navy aviators, corresponding about issues with the military's Wright aircraft, their motors and controls, and the general state of the field. Loening discussed the viability of airplanes with tractor propellers with army lieutenant Thomas DeWitt Milling, the service's first rated pilot. Loening wrote that he and Orville had "argued the matter of the tractor . . . at great length," and that they believed that a front-facing propeller was insufficient by itself to stabilize airplanes. But, Loening told Milling, it would not matter. "I understand on good authority that abroad, the tractor types are going out of favor, solely because of the difficulty of observation," he wrote, but he felt "sure that with a little study, this matter of poor observation can be overcome . . . and I am at work on a monoplane design with this in view." Suggesting that military pilots preferred the view from pusher machines, which gave their occupants unobstructed views of the terrain ahead, Loening promised to send Milling, who was then attached to the office of the chief signal officer at the War Department in Washington, a sketch of his design; he never presented the design to Orville for his consideration or for potential Wright Company production.[17]

With navy captain Washington I. Chambers, the first director of U.S. naval aviation, Loening discussed the company's seaplanes, the CH and the G, urging their purchase. Loening realized "that there is strong prejudice against our six cylinder motor, but I am confident" that a new version of the motor would allay any objections. He also wrote of the military's concerns over the nonstandard controls used by Wright airplanes (the lever system), telling Chambers that the company was experimenting with standard controls and that "Mr. Wright is entirely in favor of proceeding with this matter" but that a decision would have to be final, since "it is poor business to be changing such things from time to time." The controls the Wright Company was using as its models—from Nieuport, Deperdussin, and Breguet airplanes—were all of French origin. The two also discussed ways to make seaplanes adaptable for land use by marines. But Loening's letters did not translate into significant additional business for the company. His correspondence with his military contacts ebbed as 1914 began, with the state of the Wright Company's airplanes the likely cause: military aviators did not have questions or concerns about airplanes they weren't flying. Indeed, Loening held a leadership role in a company increasingly unpopular among military aviators after a series of fatal Model C accidents from 1912 to 1914. Oscar Brindley, who had worked as an instructor for the company at its Huffman Prairie flight school and then moved to San Diego to teach army pilots to fly Wright airplanes, informed Loening in December 1913, "One can't imagine [the] prejudice against the Wright machine and Wright motors that exists here." Army pilots, he wrote, preferred Burgess models. Moreover, Brindley told Loening, future Wright airplanes ordered by the army needed to be "thoroughly tested before leaving Dayton" or "the Wright Company's chances for any future business with the Government will be very slim." Worried for the company, Brindley simultaneously wrote a less forceful letter to Orville Wright, advising him that commanders believed "it is even dangerous for the mechanics to work on the Wright machines, let alone an officer attempting to fly one." Slim that business did become, for the army acquired only one Wright Company plane—the Model F—after Brindley's letter. Still, the military extended some courtesies, if not orders, to Dayton. With Orville Wright often unwilling to travel due to the lingering effects of his 1908 accident at Fort Myer, Loening joined other aviation leaders on USS *Mississippi* as the navy moved its aviation operations from Annapolis to Pensacola in February 1914. Loening's military contacts would prove valuable after he left Dayton.[18]

Career development and networking with prominent aviators was important to Grover Loening while he was in Dayton. He did not give the same significance to developing his personnel management skills. His relationships with the factory employees were poor. Decades later, former machinist Tom Russell (no relation to Frank Russell) remembered the young Loening as someone whom he and the other working-class "boys" in the factory disdained, since "he went around with his nose in the air." The intellectual, urbane New Yorker had little in common with the Dayton men who built the airplanes. Of course, Loening was not the first person working for the company in Dayton from an exclusive background. Frank Russell, who also came from a well-off family and whose uncle was a company director, was a member of Yale's class of 1900 who had represented U.S. interests in Canada. But Russell was thirty, several years older when he arrived in Dayton than Loening (twenty-four) had been, and Russell was married and had children. Moreover, Russell had served as an officer with managerial responsibilities at Automatic Hook and Eye; Loening, as the nonexecutive chief engineer of Queen Aeroplane. Loening's limited background in personnel management colored how the factory workers viewed him. His use of Pittsburgh's American Service Company, an employment firm, to fill a vacant plant superintendent position instead of promoting from within did not endear him to the workers. Nor did overt favoritism demonstrated to Herman Schier, an acquaintance of his from New York. Loening hired Schier, a German immigrant, to work for the Wright Company in Dayton for $24 per week, a wage of 40 cents per hour for a sixty-hour week (six ten-hour days)—2.5 cents per hour higher than the maximum rate Orville Wright had offered to a contact of Andrew Freedman's three months earlier for the same type of work. Wright did not leave any record of interfering in Loening's actions; by this time, hiring and firing the rank and file was a task he left to the man who had had such a desire to work with him.[19]

Loening had come to Dayton to work with Orville Wright, though, and not with any specific factory worker. Relationships with the laboring staff were not important to him—a recommendation from Tom Russell would not help him build his career in aviation. He placed much more importance on his relationship with Orville, but found the coinventor of the airplane to be a difficult man to work with and for. Wright's reluctance to develop the company as technologies and markets changed frustrated Loening. He kept in close contact with such army and navy officers as Milling and Chambers, trying to incorporate the results of their experiences with Wright airplanes into the company's products while he tried to make the military a reliable, ongoing customer. Nevertheless, Loening was unsuccessful in molding the company's product line to the needs of this deep-pocketed buyer. Loening also wanted to broaden the Wright Company's geographical market. He and *Flying* editor Henry Woodhouse, an Italian immigrant who had served time in a New York prison on a murder charge before establishing his magazine with the assistance of Robert Collier, developed a plan that would have sold Wright Company airplanes, and the airplanes of five other manufacturers recognized by the U.S. Army, to customers in Central and South America through one syndicate. Orville Wright would not hear of the plan. Always protective of the integrity of the 1906 patent, Wright did not want to share potential profits with other, potentially infringing companies or to have his airplanes sold alongside other makers' models. Nor did Loening have any influence over Wright's management style, which made it difficult to operate a profitable business. To Loening, Wright was "good at business" in that he was not easy to fool, but he "certainly did not have any 'big business' ideas or any great ambition to expand." Letters to the company that required Wright's attention went unanswered for weeks, and Wright avoided business travel due to his sciatica. Even though the industry was moving from pusher- to tractor-type airplanes, Loening found Wright unwilling to follow. He later wrote that "copying someone else was not in Orville's makeup," with Wright insisting that the tractor style "is really an invention of the French, and we should not be copying it just to keep up." Wright believed that propellers in the front of an airplane would interfere with the view of military observers, the "chief use of airplanes for the military," and that the "only real merit [of the tractor propeller] is that the plane can go a little faster." To Loening, Wright was "a great genius, but a troubled one," whose disaffection resulted from Wilbur's death, his chronic sciatica, and "the one great hate and obsession, the patent fight with Curtiss." Loening was not going to change Wright.[20]

Nor was Loening going to change New York. He found that Wright's relationship with the company's New York office, run since 1909 by Alpheus Barnes, who had the support of the New York directors, was both hostile and deteriorating. Loening believed that that relationship affected company operations to such an extent that it created an atmosphere of discomfort between Wright and the board that led to Wright buying out almost all the company's other stockholders in 1914. Reminiscing in 1968, Loening wrote that "it was enough for Al (who made frequent visits to Dayton and considered Wright an 'Ohio hick') faintly to suggest that something be done for Orville to place an emphatic 'no' on it." Wright returned Barnes's disregard, finding the weekly expenditure reports he demanded to be a "meaningless nuisance." Barnes tried to get Loening on his side. When Loening returned to Dayton from observing the navy's transfer of its air operations from Annapolis to Pensacola, he found Barnes in his office and began chatting with him. Orville, Loening wrote, then came into the office, made a disparaging remark about Barnes's lit cigar, and asked Loening to come later in the day to Wright's office on West Third Street. Loening recalled Wright delivering his remarks in "a clipped tone, which distinctly intimated that Barnes need not be let in on any of it." Still, Loening was not on the side of the secretary-treasurer. He found Barnes "patronizing" and manipulative, someone who was trying "to use him to spy out Orv's strange tactics and really find out where we are headed" and believed that he (Loening) "probably had become [Wright's] spy." Tired of having his ideas and designs disregarded by Wright and frustrated with being in a difficult place between Wright and Barnes, Loening left the Wright Company in July 1914 to accept a job as chief aeronautical engineer with the U.S. Army's Aviation Section at an annual salary of $3,600, writing Lieutenant Milling that "if he [Wright] is going to run the factory over my head, he certainly does not need me. He seems most unappreciative and is most decided in his views."[21]

Loening reported to San Diego to coordinate a small research and development office and to take charge or airworthiness tests on new army airplanes. He found that the army was no longer flying Wright models (it cut its two Wright Ds from its inventory the day Loening left Dayton) and was unlikely to acquire any of its new airplanes from Dayton. As Oscar Brindley had informed Loening and Wright the previous December, Wright airplanes remained unpopular among the army's pilots, and they were unlikely to recommend that the service pursue new opportunities with the Ohio company. Air force historian A. Timothy Warnock asserts that the Wright Company's responses to complaints about their airplanes' safety were "cavalier" and "alienating" to army pilots, and Loening found the army's arguments convincing. The airplanes needed to go. Loening, a decorated veteran of the aviation industry by the time he wrote *Takeoff into Greatness*, in 1968, a year before he was enshrined in the National Aviation Hall of Fame, in Dayton, developed an inflated perception of his importance in San Diego during the five decades between his departure from California and the writing of his book. He retrospectively claimed that he personally grounded the army's entire Wright and Curtiss pusher fleet for being unsafe immediately upon arriving in San Diego. While Loening certainly supported the action, the twenty-six-year-old engineer did not initiate it. Rather, Brig. Gen. George Scriven, the chief signal officer, had grounded all Wright B and C aircraft in February 1914, when Loening was still working in Dayton. A subsequent board of review composed of army lieutenants Benjamin Foulois, Townsend Dodd, Walter Taliaferro, Carlton Chapman, and Joseph Carberry—not Loening—recommended that the Signal Corps fly only tractor airplanes, an action that all but ended the Wright Company's relationship with the U.S. Army.[22]

Loening's move to the army again left the Wright Company with a vacant leadership position, one it would not fill during the remainder of Orville Wright's ownership. Instead, as had happened when Frank Russell left, existing employees assumed some of his former duties. Carl Nellis, of whom Loening thought highly enough that he attempted to entice him to move to San Diego to work for the army, informed Loening that he had "full charge of the Wright Shop" as a superintendent in late 1914, with "very pleasant" wages. Meanwhile, Wright now had undivided managerial responsibility, an adjudicated patent (which had little effect on Loening's work; he was not involved in the company's legal battles), and soon all but complete ownership. His plans for the company no longer involved making it the nation's monopoly airplane manufacturer. However, until he obtained majority ownership, Wright still had to deal with the New York face of the company: Alpheus Barnes.[23]

Entwined between Frank Russell, Grover Loening, and Orville Wright was Alpheus Fayette Barnes of Jersey City, New Jersey, who as corporate secretary and treasurer ran the New York office and periodically came to Dayton to monitor factory operations (and, according to Grover Loening, Orville Wright) on behalf of the New York–based directors, especially after Wilbur Wright's death. The son of a New Jersey clerk who died when his son was two, Barnes is a difficult man to study. His route to the Wright Company is cloudy, and his life afterward is mysterious. He never gained personal prominence in the aviation industry, but as the representative of the New Yorkers who invested in the Wright Company, he attempted to mold the company into the profitable airplane monopoly they had envisioned when they contributed their capital, in 1909. The New York investors viewed Barnes as their agent, not their equal. They did not leave their impressions of him—indeed, very few of the directors left any impressions of the Wright Company at all. However, the men with whom he interacted in Dayton viewed him as the embodiment of those investors and left vignettes of Barnes in their writings.[24]

The loquacious Grover Loening provided the most extensive commentary on Barnes and his work. Loening's recollections of Barnes evolved over the three decades between his books. Writing in *Our Wings Grow Faster*, in 1935, he described Barnes as "a hearty, well-built . . . genial character, smoking cigars continually and full of good stories" who "did all the bookkeeping work as well as advertising and negotiating of contracts, and left the management of the plant pretty much up to me." Thirty years later, in *Takeoff into Greatness*, Loening remembered Barnes as a "patronizing, somewhat fatherly" figure who, on behalf of the board, wanted to get Loening to "induce Orville to get out some newer and more modern designs than the *Model B* then in production," yet who "was not very high in Orville's esteem, nor did he have much influence in the factory's business operations." Orville felt the same way about Barnes and, Loening recalled, was rather dubious about the New York directors generally. Though they were prominent, successful capitalists, he thought they had no clue on how to run an aviation business, applying "old manufacturing rules and customs to something that is so different" in its operations, he thought, than other industries. Wright, believing he knew best, frequently ignored official correspondence from Barnes, which ironically caused the New Yorker to make inspection visits to Dayton, especially after the 1914 patent lawsuit victory, when the directors wanted him to help create an industrial monopoly that, Loening recalled Barnes exclaiming, would be "a damn sight bigger than any of these dumb Ohio clucks can even imagine." The secretary Loening remembered was a brash man convinced of the superiority of his and his patrons' experiences to those of Wright, full of ideas Wright did not want to hear.[25]

Loening, of course, had no experience of Barnes's work when Wilbur Wright was company president, a situation to which Frank Russell was privy. Though Russell and Barnes usually communicated by letter or wire (some of Barnes's correspondence to Russell survives, but Russell's side of the correspondence generally does not), Russell, like Loening, personally interacted with the cigar-chomping New Yorker. According to Russell, Barnes was critical of the firm's business operations from its earliest days. The new general manager first met with Barnes and the other directors in New York on 5 January 1910 before moving to snowy Dayton to take up his new job, noting in his diary that Barnes was already dissatisfied with company operations. The newly ensconced secretary believed he was uninformed as to what was going on (this before operations had even started in Dayton; Russell's hiring was what was "going on") and "wished [the] factory was in the east." Perhaps Barnes hoped that Russell would be his spy. Later that winter, Russell hosted Barnes for lunch at the Yale Club in New York. There they spoke about the goals of the New York investors, and Barnes promised to keep Russell informed of New York operations, anticipating that Russell would do the same from Dayton. Barnes also watched over Roy Knabenshue and the exhibition department. With Knabenshue often on the road with his aviators, Wilbur Wright requested in July 1910 that Barnes come west to "take charge of an office here in the absence of Mr. Knabenshue and to supervise the collection of the exhibition accounts" for "a few months." Vice president Andrew Freedman approved Wright's request, noting that "it would be perfectly satisfactory" to have Barnes working in Ohio.[26] Freedman himself did not have the best relationship with Barnes; accompanying Russell on the train to Dayton after the exhibition department's first appearance, in Indianapolis in June 1910, Barnes told the general manager that he considered Freedman a dishonest, power-hungry man, a "grafter like the rest" of the directors. Barnes's relationship with Orville Wright was poor even before Wilbur Wright's death. Russell noted a conversation with the younger brother in July 1911 where Wright stated that the "N.Y. situation *can't be helped* with Barnes there. Holding such a low opinion of the secretary, the Wrights avoided socializing with Barnes when he came to Dayton. They dined with him much less frequently than with Russell, inviting him to their homes on only four occasions during his many trips to Ohio between 1910 and 1914. The Russells were fairly frequent guests at 7 Hawthorn, and Grover Loening had the privilege of dining atop the widow's walk of Hawthorn Hill while it was still under construction, on "the hottest day of the very hot summer," 30 July 1913; he escorted Katharine Wright to "some evening show" a few days later and made frequent calls (though not enough for Milton Wright to learn to spell his name) during his year in town.[27]

FIGURE 8.3. Alpheus Barnes at work. *System: The Magazine of Business* 17, no. 2 (February 1910): 218.

The brothers usually joined their father and sister, Katharine Wright, for supper when they were at home; people connected with the company or aviation, as well as other family members (especially their elder brother Lorin and his family), were also often at the table. However, during the two and a half years of Wilbur Wright's presidency, the brothers spent significant amounts of time apart from each other, pursuing the Curtiss lawsuit in New York and business affairs in Europe. During their separations, they corresponded frequently, discussing technical aspects of their airplanes and the operations of their European and U.S. companies, as well as their nieces and nephews and plans for the house that became Hawthorn Hill. Much was going on in between 1910 and 1912—from the creation of the Boy Scouts of America (for which Orville would later help create an aviation merit badge) and Girl Scouts of the U.S.A. to the breakup of John D. Rockefeller's Standard Oil, the sinking of the *Titanic* in the North Atlantic, and Theodore Roosevelt's return to politics in what would be his Bull Moose campaign for the White House. But these things did not interest the Wrights. Politics, women's rights, or racial or labor issues rarely gained any of the brothers' ink. Nor did Alpheus Barnes. He did not provoke the brothers in those years to the extent that he later provoked Orville, and he rarely appeared in their letters. Their occasional mentions of him are perfunctory, noting his monitoring an exhibition department appearance in Hartford (where Barnes was "disgusted" by Clifford Turpin's flying—Turpin had raced against an automobile at Charter Oak Park, outdistancing it) or that he was staffing the Wright Company shed at the 1911 Nassau meet in New York, or pursuing Cal Rodgers's *Vin Fiz* expedition while he traveled through New York to collect an unpaid license fee. While Barnes was a member of the company's board of directors, the Wrights ignored his executive role (and the role of the other directors). Instead, their discussions of company management usually addressed—and rather critically—the work of Frank Russell or Roy Knabenshue. Recognizing that Barnes had the support of the directors in New York, neither Wright ever attempted to remove him from his secretary-treasurer position while they shared company ownership. Once Orville Wright acquired majority control of the company, in the autumn of 1914, though, Barnes's role receded. Orville took no action to formally remove him from office but instead ignored him entirely. Barnes remained an officer until the last, being replaced by Katharine Wright when Orville sold the company in October 1915. Barnes, Katharine noted in her meeting minutes, was "no longer in the employ of the company nor in any way attending to the company's affairs."[28]

Uncovering the person who was Alpheus F. Barnes is difficult work. He left no collection of personal papers, and the official correspondence he maintained with Russell, Loening, and the Wrights discusses company operations—audits, financial procedures, the financial success or failure of exhibition appearances. His correspondence is incomplete and contains little substantive discussion about the patent lawsuits, technological development, or corporate policy, or about Barnes himself. Once rid of Barnes, Orville Wright never corresponded with him. Neither Frank Russell's nor Grover Loening's papers contain evidence that Barnes played any further role in the U.S. aviation industry, and he apparently did not have time to have much of an effect anyway. With his wife, Sarah Barnes, listed as a widow in the 1920 U.S. census for Jefferson Township in Morris County, New Jersey, Alpheus Barnes never had the opportunity to develop a career in aviation like Frank Russell or Grover Loening, even if he wanted to do so, though his son, Alpheus F. Barnes, Jr., listed his occupation as "pilot" on his 1918 draft registration card. But by the time the 1920 census taker came around, the man Russell, Loening, and Barnes had worked for had also forsaken an active career in aviation. Orville Wright had retired to Hawthorn Hill, to serve on commissions and boards, experiment in his west Dayton laboratory, and dote on his nieces and nephews. During the year in which the Aero Club of Dayton elected him an honorary member and his elder brother Reuchlin died (and during which women gained the right to vote but lost the right to drink), Orville Wright reached the fifth anniversary of his sale of the Wright Company.[29]

9

# "It Is Something I Have Wanted to Do for Many Months"

*Exit Orville*

Much had changed in the world and in Orville Wright's life between 1909 and 1915. Europe was no longer a welcoming destination; trenches filled with soldiers divided Germany from Belgium and France. Ohio Republican William Howard Taft had been replaced in the White House by New Jersey Democrat Woodrow Wilson. After nearly a century without, the United States again had a central bank in the Federal Reserve. The city of Dayton had been through a devastating flood and adopted a new form of municipal government, with a powerful city manager and a weak mayor and council. Wilbur, his brother and partner in inventing the airplane and in running the Wright Company, had been dead for three years in 1915. Orville had moved with his father and sister from the small house of his birth, on a cramped lot in west Dayton, to an expansive mansion on a tree-covered hillside in the city's southern suburbs and had become the sole owner of a company first capitalized at $1,000,000. Aviation had also changed, and not in ways favorable to Orville's conservative approach to corporate development. Changes in airplane design and construction in Europe and the United States, often undertaken by the Wrights' competitors to try to avoid patent lawsuits, had overtaken the work of the surviving father of the airplane. In Europe, generous government allocations enabled companies in Germany, the United Kingdom, and (especially) France to develop aircraft that technologically surpassed those built by the Wright Company and its North American competitors; they were easier to fly and faster and safer in the air. And the situation on the western shores of the Atlantic had also changed. In New York, Glenn Curtiss no longer considered the Wright Company a significant competitor. His company in 1915 was much larger and more successful than it had been in 1909, building many more airplanes than the Wright Company. Curtiss's success enabled it to build new factories and purchase the Burgess Company in 1916. In California, the young Glenn Martin's works grew from nothing to an enterprise with which Orville Wright's successors as Wright Company owners would combine the former Dayton concern in 1917. Wright himself had been a leading aviator and airplane designer in 1908 and 1909, years when he and his brother thrilled royalty from across Europe and dignitaries in the United States with their airplanes, gaining gold medals from Congress, the state of Ohio, and the city of Dayton, which held a two-day celebration of them and their achievements. By 1915, though, Orville rarely flew or created designs that affected the industry. Over the six years he had become an elder statesman of aeronautics. He still had a respected name and, through his company, a valuable patent, but he no longer had his brother and business and creative partner. The Wright brothers were now a part of history; Orville Wright was the comfortable owner of a company that no longer built airplanes that people wanted to buy.[1]

His personal finances secure, Orville Wright was reluctant to remain in the business world. In the years after Wilbur Wright's death, in 1912, Orville, proud of his and his brother's invention and reluctant to see it changed, did not try to increase corporate profits by answering changes in the market for airplanes. His resistance to incorporating significant design changes into Wright Company aircraft—to installing a more intuitive control system, to

FIGURE 9.1. Orville and Katharine Wright in a Model HS, 1915. *Courtesy of the Library of Congress*

ailerons instead of wing warping, to tractor-style airplanes—changes that might increase the market share of the company's products and the profits accruing to its stockholders, contributed to the company's decline, and to an unhappy board of directors. Wright was already at odds with Alpheus Barnes over routine management issues, but his preferred course of corporate action after the successful conclusion of the patent lawsuit against Curtiss, in January 1914, worsened his relationships with the Wall Street financiers who had invested in the company and served on its board of directors. After the U.S. court of appeals affirmed Judge Hazel's district court decision in favor of the Wright Company, Wright and the board met in New York to discuss how to capitalize on the ruling. Most of the board members, including "master of transportation" Theodore Shonts, vice president Andrew Freedman, and Barnes, wanted to use the opportunity to aggressively market the company's wares to the U.S. government, which had not yet stopped acquiring the pusher-style airplanes produced in Dayton and, they believed, would certainly not want to purchase illegally built airplanes. But Shonts, Freedman, and most of the other directors lived in New York, and the Wright Company was just one of the many concerns in which they were involved (and it was a small concern when compared with the Interborough Rapid Transit Company). No one connected with the Wright Company spent much time where the government purse resided: Washington, D.C. Therefore, the directors believed that the company needed to hire someone well connected with the levers of federal power to lobby the halls of Congress for sales, and they suggested for the job a lawyer whom many of them knew well: William F. McCombs. McCombs (1875–1921) was a New York lawyer, the chairman of the Democratic National Committee, and a confidant of Woodrow Wilson who managed his successful campaign for the presidency in 1912 and, like Andrew Freedman, was closely connected with Tammany Hall machine politics in New York. Wright reacted poorly to this suggestion, and his reaction sparked heated discussion. Shonts, Freedman, and Barnes and a majority of the other members of the board supported setting an aggressive policy of suing Curtiss and other infringers, forcing them out of business, and giving the Wright Company a monopoly in the industry. They wanted someone like McCombs, who had deep connections to the executive branch, to represent the company in Washington and lobby for sales to the federal government. President Wright vigorously dissented; he would not have the Wright name represented by a machine politician. He "was perfectly willing to retain some other good lawyer if the Board desired it" to lobby, but he did not want his name to "be mixed up in politics of this kind." Wright stood alone in his desire to keep politics out of the boardroom; he was president of a company with a board of directors composed of men with deep political connections, especially with the Democratic Party. Wright vice president Andrew Freedman had been the national treasurer of the Democratic Party in 1897; De Lancey Nicoll served as Democratic National Committee vice-chair in 1907; and August Belmont was the son of August Belmont, Sr., chairman of the (Northern) DNC during the Civil War and was instrumental in the nomination of Alton B. Parker for president of the United States in 1904. Meanwhile, Orville Wright was a man from a very different background. He was not politically dogmatic—Wright supported a variety of politicians from different parties during his life, marching in a parade in honor of president-elect Benjamin Harrison as a teenager and, according to his father, voting a straight Democratic ticket in 1916. But he came from a moderate Republican family with no substantial political connections or inherited wealth, and he had little personal interest in active politics. Even Republicans Russell Alger, whose father had served as governor and U.S. senator from Michigan and as the first secretary of war in President William McKinley's administration, and Theodore Shonts, who served in the administration of President Theodore Roosevelt as chairman of the Isthmian Canal Commission in Panama from 1905 to 1907, recognized the wisdom of having a Democrat approach a Democratic administration and supported McCombs's engagement. But the directors' world was different than Wright's. While Wilson's immediate predecessor in the White House, Republican William Howard Taft, was from nearby Cincinnati, Wright had no connections with the former president and future chief justice. Nor did he have connections to the inner workings of Tammany Hall or to national politicians maintained by most of his company's officers, or to any sort of a Republican political machine in Dayton or beyond. Wright did not even make any personal contributions to a party; instead, his politically oriented donations supported Prohibition (especially the Anti-Saloon League) and women's suffrage. He did not want the public to associate the Wright name, or aviation as a whole, with a particular political ideology. Wright's opposition to the board's proposal was sufficiently strong to keep McCombs from representing the company, which never employed anyone to be its advocate in either Congress or the White House.[2]

The disagreement over McCombs was also enough to permanently damage Wright's already shaky relationship with the board. Shonts, Nicoll, and Freedman were not happy with what they viewed as Wright's obstinacy toward an action that they expected would greatly increase the company's revenues. Of course they wanted to hire a prominent Democrat to lobby the administration of Woodrow Wilson, a Democrat. Moreover, Democrats controlled both houses of the Sixty-Third Congress by wide margins. A Republican-oriented lobbyist would find it difficult to gain much attention from those holding the levers of power. The board members argued that "all of the people who had invested in the company would lose their money" if Mc-Combs was not retained to lobby Congress, the White House, and the War Department to purchase airplanes from the Wright Company. In response, Wright made a fateful decision for the Wright Company's future. Eager to take full personal control of—and responsibility for—the company, Wright proposed to ensure that no director would lose the $20,000 investment he made in 1909 ($450,000 in 2010 dollars). He "offered to take their stock off their hands at a price . . . [that] would give them back all the money they had invested with ten per cent interest per year." Buying all the stock would essentially (though not legally) make the Wright Company a sole proprietorship under Wright's personal control. After conferring among themselves, Nicoll, Freedman, and Shonts agreed to sell their stock to Wright at its original price if he agreed to make the same proposal to the board members who were not present and "take all of the other stock that is offered at the same price." Wright soon made them former investors in his work. He further explained his decision in a letter to Frederick Alger, the younger brother of director Russell Alger. By being the owner of all the company's stock, he wrote, he would be able "to grant licenses without having to consider any one's interest excepting my own," would not have to share any profits, and, more important, would "be in a position to accept a cash offer for the entire business . . . and thus be relieved of all responsibility." No longer a bicycle maker of moderate means, Wright financed the purchases, which totaled nearly $200,000, by dipping into his checking account at Dayton's Winters National Bank and by selling some municipal bonds he owned. By June 1914, a "delighly" cool month in Dayton according to Milton Wright, Orville Wright controlled all the executive offices—having sidelined Alpheus Barnes from his position as company secretary—and all the stock but Robert Collier's 240 shares (3 percent of those outstanding). Wright did try to purchase the stock from Collier, but during much of 1914 Collier was either traveling overseas or ill and did not make arrangements to sell his shares to Wright. Working through lawyer Pliny W. Williamson, Wright was in no hurry to acquire Collier's shares. According to Grover Loening, Collier was "a close and loyal friend of" Orville's, and with just 3 percent of the shares he was not in a position to dictate corporate policy. Wright even wrote sympathetically to him, telling Collier that he would have been "very glad to have you remain a stockholder in the Company" if he had not wanted to sell—and he needed 100 percent control to sell.

Williamson, an acquaintance, west Dayton native, and Oberlin College classmate of Katharine Wright who considered himself part of New York's "Dayton Colony," purchased Clinton Peterkin's stock in 1910. For several years, Williamson had sent the Wrights chatty letters with clippings from New York newspapers about the company, its competitors, and aviation. Williamson entered politics in 1912, becoming Republican chairman in the town of Scarsdale. In 1929 he was elected to the Westchester County Board of Supervisors, and in 1934 to the New York state senate. He served in Albany until his death, in 1958, and was known there as a respected legislator who served his colleagues boxes of raisins instead of the traditional highballs at the end of a legislative session. But in 1914 he was handling the company's nonpatent legal affairs in New York and informed Collier's lawyer that the matter of the stock was something that could wait until Collier and Wright met at some future time. Through his attorney, Collier eventually told Wright that he would retain his stock as long as

Wright owned his. Ultimately, Collier's 3 percent ownership, which he retained until Wright sold the company, did not matter. With 97 percent of the shares under his control, it was truly Wright's company to do with as he pleased.[3]

But it was still not officially a Dayton company. The Wright Company remained a corporation formally registered in New York, one that was required to have individuals appointed to its executive offices, if only as a formality. Orville Wright filled those offices, but not with the titans of Wall Street who had invested in 1909 and 1910. Instead, he looked to Dayton and to family for new board members. Orville's elder brother Lorin Wright (1862–1939) replaced Freedman as vice president in 1914. Lorin, though never a Wright Company employee, served as a spy for the firm. In 1914, Glenn Curtiss, looking for another way to get around the Wright patent after losing decisively in court, remembered that Smithsonian Institution secretary Samuel Langley had attempted to launch his Aerodrome skyward from the Potomac River a few days before the Wrights' successful flights at Kill Devil Hills in 1903. Langley's attempt to fly failed, but Curtiss wanted to see if Langley's airplane was capable of flight under more controlled conditions. More than a decade had passed, and there were now many pilots more familiar with how to control an airplane than was Samuel Langley's pilot, and perhaps the courts would invalidate the Wright patent if Curtiss could show that someone else was capable of flying before the brothers but had not actually done so. Langley had died in 1906, but Smithsonian secretary Charles Walcott agreed to take the Aerodrome off exhibit and loan it to Curtiss for testing. In 1914 he shipped the Aerodrome to New York, where Curtiss announced he would restore it to its 1903 condition for the tests. Curtiss actually made several major modifications to the Aerodrome that made it much more airworthy than in 1903, strengthening its wings especially, and he successfully flew it a short distance on Keuka Lake in May. Curtiss added a new engine, and continued to make modifications to the Aerodrome through 1914 and early 1915. Orville Wright, having seen photographs of the "restored" aircraft taken by his friend Griffith Brewer, a British aviator and patent attorney, was outraged, and in the summer of 1915 he sent Lorin to New York to investigate. The balding, mustachioed Lorin, a bookkeeper by profession, was perhaps not the best spy for the balding, mustachioed, well-recognized Orville to send, but he went willingly to protect his brothers' honor. He first went to Toronto to see what the Curtiss operation there was doing. Wiring his brother from the King Edward Hotel in Toronto, where he had inspected Curtiss hangars on the Toronto Islands, he noted that Curtiss's Canadian company operated the facility, but that "everything including doors and fittings built Buffalo." The next day, he visited Curtiss's Toronto factory, where he noted nearly eighty men at work and two airplanes on the shop floor; he then boarded a night train for Buffalo to reach Hammondsport the next afternoon "to see what they are doing." He obtained a hotel room in Bath, seven miles outside Hammondsport, under the name of "W. L. Oren," and proceeded to poke around the Keuka Lake hangars, taking lots of photographs. Though Curtiss's associates confiscated his film, Lorin remained observant and sent Orville a detailed letter identifying some of the changes to the Aerodrome—rear posts, the rudder, the floats. In the summer of 1915, Orville was not interested in suing Curtiss again; he was working on selling the company. Instead, Lorin's work—and the Smithsonian's reinstallation of the Aerodrome with a label claiming that it was truly the first airplane capable of flight—set off a long fight between Orville and the museum over the ultimate disposition of the Wrights' 1903 airplane. Furious at the inaccurate text of the label, Wright sent the 1903 airplane to London, where it was a prime attraction at the Science Museum from 1925 until it was removed ahead of German air raids during the Battle of Britain. Only after the Smithsonian agreed to relabel the Aerodrome did the 1903 airplane return to the United States for exhibition, shortly after Wright's 1948 death.[4]

Lorin was the most active of Orville's siblings on company business in 1914 and 1915, but Katharine Wright also became a Wright Company officer (their eldest brother, Reuchlin, lived in Kansas and had no involvement with the company). Katharine and Estelle L. Beatty (no relation to aviator George W. Beatty, she was the wife of a Dayton brass molder who had worked as one of the company's office secretaries at the factory) both briefly occupied the secretary and treasurer offices in October 1915 as Wright sold the company, Katharine for a day and Beatty for six. Wright first proposed that his brother and sister join the board in a November 1914 telegram to Pliny Williamson. Not specifying a particular office for either woman, Wright merely suggested that they and Williamson join Wright and Alpheus Barnes, who was not formally relieved of his office until just before the company's sale, in 1915, as directors. The incompleteness of company records makes it difficult to determine whether the new board was seated before October 1915 or whether Wright's suggestion, though agreeable to Williamson, remained unimplemented. Only in office to satisfy legal requirements, once on the board, neither of the Wright siblings nor Estelle Beatty impacted company operations or policy.[5]

Orville Wright remained a prominent aviation figurehead for the rest of his life, accepting honors and serving on boards, but few of the company's other former directors maintained significant connections with the field. Robert Collier gave the most attention to the fledgling discipline, serving as the president of the Aero Club of America from 1911 until his death in 1918. Andrew Freedman died shortly after the sale, remembered for his role in building New York's subway system, not for his connections with the Wrights. When Theodore Shonts died, in 1919, his obituary also focused on his time working for railroads and his role in the development of the Panama Canal; the Wright Company was not mentioned. Russell Alger was memorialized as a Packard officer, while De Lancey Nicoll gained attention as a former district attorney of New York City and lawyer for publisher Joseph Pulitzer. As with Freedman and Shonts, the roles of these men in the affairs of the Wright Company were not deemed noteworthy by their obituary writers. As secretary and treasurer and the New York face of the firm, Alpheus F. Barnes was much more active in company affairs than any other board member save the Wrights themselves. Barnes, though, disappears from the record after leaving the Wright Company. He likely died within the next several years, since his wife, Sarah Furgo Barnes, is listed in the 1920 U.S. census as a widow, but the New York newspapers that detail the deaths of the other board members include no death notices for Barnes between 1915 and 1920. The life of one of the more influential Wright Company directors in the years after the company left Dayton remains a mystery.[6]

What happened to the Wright Company's workers? Without the socio-economic prominence of a Vanderbilt or Freedman, their lives attracted little attention from the press, and none donated their personal papers to an archive. Cemetery records, obituaries, census returns, and, in a few cases, interviews conducted at the University of Dayton in the late 1960s are the best guides to their stories. Most took other industrial jobs in Dayton or southwestern Ohio after the Wright Company relocated; no one seems to have followed the firm to New Jersey. Only Charles Taylor, who gained a measure of fame through his role in building the engines of the Wrights' first airplanes, received much attention from the press in his later years; his financial difficulties and death at the age of eighty-seven, in 1956, gained the attention of the Los Angeles Times, and the Associated Press announced his death across the country. Taylor's fame grew after his death. In 1965 he was enshrined in the National Aviation Hall of Fame, and the Federal Aviation Administration created the Charles Taylor Master Mechanic Award to recognize those who worked for fifty years as certified aviation mechanics in 1993. Several of the company's other workers acquired enough wealth to purchase their own homes, while others remained renters. Census takers, recording data six months after the October 1929 stock market crash, show that Arthur Gaible moved to Cincinnati, where he worked in the construction industry and rented a house for his large family. Rufus B. Jones also moved to the Queen City, where he became a mechanical engineer for a company that made trailers for trucks. Jones, in 1930, owned a house worth $7,000. Meanwhile, Louis Luneke remained in Dayton, where in 1930 he was working as a mechanical engineer for Frigidaire and owned (with his wife) a house worth $7,500. Frank T. Whipp, who worked for the Wright Company as a woodworker, and who in 1930 owned a house in Dayton worth $5,500, still worked in the aviation industry as an airplane builder, more than a decade after the Wright Company left town. Luneke and Whipp remained employed and in their houses throughout the Depression.[7]

Tom Russell, William Conover, and Fred Kreusch were among the participants in the Wright Brothers–Charles F. Kettering oral history project sponsored by the University of Dayton in 1967. Though retired at the time, all three had enjoyed successful careers with Dayton manufacturers. Meanwhile, Robinson Elliot, who worked for the Wright Company between 1909 and 1912 and retired from Delco in 1948, was unavailable for Susan Bennet to interview, having died at the age of ninety-one, in 1963. Elliot received a brief obituary in the *Dayton Daily News* that stated he "was instrumental in having a punch press installed in the plant to produce parts formerly made by hand, and constructed the die used in the machine." The *Daily News* carried death notices for Whipp, Luneke, Russell, and seamstress Ida Holdgreve, who, at the time of her death, at the age of ninety-five, in September 1977, was the last surviving employee of Orville Wright's Dayton company.[8]

Of course, none of these workers ever owned a piece of the Wright Company. By the middle of 1914, it was Wright's personal property, to operate as he saw fit. He did not see fit to attempt to reopen sales to the U.S. military or to try to reach contracts to supply airplanes to European powers, as did Curtiss with the Russian Empire, though the company's New York office did receive inquiries asking about its "willingness to undertake foreign business." Those inquiries the company turned away; Wright believed that his company's engines were not powerful enough for military use in the skies of Europe. As a result, the company was not part of the significant growth in airplane exports that occurred between 1914, when U.S. manufacturers shipped forty aircraft overseas, and 1915, when they exported 398 (and exports only grew thereafter). It was not even able to avoid more patent litigation with Curtiss, who began looking for a new way to evade the strictures of the 1906

Wright patent in the wake of the Wright Company's court victory. Curtiss hired patent attorney W. Benton Crisp, who also represented Henry Ford, and on Crisp's advice changed the ailerons on his airplanes from dual control to individual control, a change that Crisp believed had not been adjudicated in court. The Wright Company could not let this stand. In November 1914, as voters directly elected U.S. senators for the first time (and as Dayton's James Cox lost his bid for reelection as Ohio's governor), it sued in another attempt to stop Curtiss, who decided to rebut the suit by challenging the validity of the 1906 patent—by showing that the Wright airplane was not the first machine *capable* of flight, and therefore not protected. That challenge led to Lorin Wright's visit to Hammondsport and the longstanding Wright battle with the Smithsonian. The 1914 lawsuit transferred to the Wright Company's new owners in 1915, and remained unresolved when the U.S. government created an airplane patent pool in 1917, stopping a court case in which Orville Wright no longer had a legal interest.[9]

Curtiss continued to build new models of airplanes in Buffalo and Toronto, supplying the U.S. and British militaries with such models as the 1915 JN-4 Jenny, famously used as a trainer during the First World War and by barnstormers in the 1920s. The Wright Company's contemporary new models—the H, HS, and K—were afterthoughts, if thought about at all. In fact, the Wright Company was still producing Model B airplanes, the company's best-selling model of 1910 and 1911. Shortly after he sold the company, Orville Wright sent a letter to Roy Knabenshue (by then working principally with dirigibles) to tell him that four of those antiquated aircraft that Knabenshue had ordered were under construction. Wright assumed Knabenshue wanted them for use as trainers and the Dayton workers, therefore, were using some recycled parts; Wright did not think that new parts were essential to complete Knabenshue's order. Wright realized that the company's airplanes were no longer of much commercial value. He also knew, though, that the Wright Company did control two valuable assets: the Wright name and the 1906 patent. With plenty of money in the bank, and eager to leave the life of a corporate president behind for a true return to experimental work, Wright—though he initially denied doing so to the Dayton press—began to entertain offers from others to purchase the company or to completely license its rights to another entrepreneur.[10]

Wright did not necessarily believe that he had to sell the company outright, though given his work with Frank Russell and Grover Loening, one wonders whether he would have been able to avoid attempting to micromanage

FIGURE 9.2. Roy Knabenshue in the basket of his *California Arrow* airship at the St. Louis World's Fair, 1905. *Courtesy of the Library of Congress*

a licensee. And he let others act as his agents. For much of late 1914 and early 1915, Wright, through Pliny Williamson, attempted to negotiate either a licensing agreement that would end company production in Dayton but keep company ownership in Wright's hands, or a sale of the business, to Samuel C. Morehouse of New Haven's Connecticut Aircraft Company. While Morehouse, an attorney, was the Dayton company's most promising suitor that autumn and winter, Franklin A. Seiberling of the Goodyear Tire and Rubber Company of Akron, Ohio, also expressed interest in acquiring the firm to complement Goodyear's interest in airships. Meanwhile, William Hammer, an electrical engineer and associate of Thomas Edison who served as a Wright

Company consultant in court cases, tried to interest Edwin Rice, the president of General Electric, in acquiring the airplane builder, but Rice declined the opportunity. After months of letters back and forth and negotiations, the deal with Morehouse died. Morehouse was a defendant in a lawsuit concerning monopolistic practices of the New York, New Haven and Hartford Railroad, which damaged his ability to raise money for a purchase, as did the start of the war in Europe. Becoming impatient with Morehouse, and reminiscent of his and his brother's insistence in the mid-1900s on obtaining down payments before showing their airplane to potential buyers, in October Wright insisted that Morehouse show a "substantial interest in the negotiations" by providing a down payment of 10 percent of an agreed purchase price on an option to purchase the company, or Wright would consider offers from other parties. The only semiserious suitor that fall, aside from Morehouse, was New York aviation publisher Henry Woodhouse. Woodhouse, who had no waiting production facility, visited Dayton to discuss licensing options in September 1914, supping with the Wright family, but went home with nothing. Wright, though, used Woodhouse's supposed interest to try to provoke Morehouse to finalize a deal, bringing him and Williamson to Hawthorn Hill for more negotiations and dinner in November 1914. But the Connecticut lawyer continued to stall.[11]

A few days after Morehouse returned to Connecticut, Wright, through Williamson, gave Morehouse his terms for the Wright Company: he would sell the company and its assets for $737,500 (nearly $16.6 million in 2010 dollars). Wright no longer wanted a 10 percent down payment—he wanted $350,000 in hand before releasing the company's stock to Morehouse. Still financially insecure, Morehouse thought he could obtain the money if the Connecticut Aircraft Company secured an order from the War Department for one hundred airplanes. It did not. Washington looked elsewhere for aircraft, and at the end of March 1915, Morehouse placed a new offer before Wright. He no longer wanted to buy the company. Instead, he proposed purchasing an exclusive production license from Wright at an unspecified price, one that Connecticut Aircraft would fund with the proceeds if it won a bid to build seaplanes for the U.S. Navy. Morehouse had better luck with the navy. While the Burgess Company wound up winning the bid for seaplanes, Morehouse's company succeeded in winning a separate bid for the navy's first dirigible, which it introduced to the public with aviator Thomas Baldwin as pilot in August 1915. But this was not enough for Orville Wright. He was tired of dealing with Morehouse and was cool to the new proposal. He would consider granting Morehouse a license, but not "indiscriminately," and he had "no very great confidence in the seriousness of the Connecticut Aircraft Company" and did not want to spend money on talks with Morehouse or other representatives. Negotiations between the parties continued for a few more months, with Morehouse still unable to assemble investors or obtain cash. In late May, Pliny Williamson informed him that it was "high time either to close it or end negotiations." Morehouse believed that the Wright Company no longer was worth as much as it was when he first approached Wright about a purchase, asserting to Williamson that its position was "not as strong as it was a year ago" but that he could restore its value through "vigorous action" of enforcing "all rights under the patents." Still, he continued to attempt to convince Wright of his ongoing interest in acquiring some piece of the company, and he and Williamson dined again at Hawthorn Hill with the Wright family on 9 June. While Morehouse remained interested in consummating a deal, and Wright, ready to conclude one, began to get his papers together, Morehouse still could not raise the cash Wright required—and he was finally out of time. By late July the Wright Company had a new suitor with plenty of cash on hand, and Orville acted quickly to conclude a deal.[12]

Even if the company's airplanes were old technology, the Wright name and patent (which would not expire for another eight years) retained value as the First World War entered its second year, and some New York businessmen with deep pockets thought they might be able to revitalize the company. From his Manhattan office, Williamson began negotiations with Frederick Y. Robertson of the U.S. Metals Refining Company and attorney Phillip W. Russell. They announced to the public that they wanted to "develop the aeroplane on a scale to make it possible for any man to own and use one and to make the aeroplane as common as the automobile." Talks proceeded rapidly in August and September, a period Williamson told Wright was the "harvest season for the aeroplane" as a result of the war. He visited Dayton several times that summer and fall for consultations, and Wright traveled to New York three times between July and October in spite of his sciatica. Robertson and Phillip Russell also met with the Wright Company's New York patent attorney, Frederick Fish, "to get exact information on [the] patent situation" and the court case against Curtiss. With the syndicate ready by the end of August to sign a contract with Wright, attention turned to Robert Collier and his stock. While Collier had told Wright that he "would do whatever [Wright] asked him to do," the "Eastern parties" were concerned about the possibility, however remote, that the publisher might become a dissenting stockholder and a thorn in the sales process. Wright eventually sent Collier a note asking for a meeting to discuss the sale, but nothing came of it; Williamson, frustrated after a month of correspondence on the topic, wrote both Katharine and Orville Wright letters asking them to at least send Phillip Russell a letter describing the situation. Over the next two weeks, Williamson and Wright mollified the concerns of Robertson, Russell, and the other members of their syndicate, and on a chilly 13 October 1915 ownership of the Wright Company passed to a group led by mining engineer and financier William B. Thompson, Chase National Bank president Albert H. Wiggin, and T. Frank Manville, an executive with the asbestos manufacturer H. W. Johns-Manville Company, as well as Russell and Robertson. For the first time in more than fifteen years, Orville Wright no longer had an active role in bringing aviation to the world. Instead, he "leaned back in his chair and closed his eyes," telling a [New York Times] reporter, "It is true I have sold the control in the Wright Company to New York capitalists. It is something I have wanted to do for many months."[13]

Valuing Wright's status as one of the fathers of the practical airplane and his institutional memory, Thompson, Wiggin, and Manville believed that the Wright Company would benefit by retaining its founder in a titular capacity. The terms of the sale provided Wright a $25,000 salary for services as a consulting engineer "relieved of all responsibility" during the first year of the new ownership's management. The syndicate had lots to do to revitalize the company. While Wright and Williamson concluded their negotiations with the New York group, Aerial Age Weekly reported that the Curtiss, Burgess, Glenn Martin, and Thomas Brothers companies were collectively shipping fifteen airplanes a day to the war. The Wright Company? Zero. Emphasizing the positive attributes of the purchase, the new owners boasted to the press that the company held valuable patents and that they intended that "development work will be pushed vigorously, and a great deal of money spent for experimental work and enlarging capacity" to enable the Wright Company to regain a place among the major U.S. airplane makers. No one in Dayton accused Wright of selling out. The Dayton Herald viewed the sale positively and expected "that the new firm will go after big orders." The Dayton Journal also looked favorably on the sale, claiming that it was the result of Wright's ill health over the previous year. The paper also felt that new management would bring additional capital to the firm and that the west Dayton factory would "engage in the manufacture of flying machines for the warring nations on a far larger scale than has hitherto been attempted, and may consequently become the largest aeroplane factory in the world." This optimism for a company that had not physically expanded its factory since 1911 or introduced a popular new model since the B of 1910 and 1911, a company that had just been sold by a man who stated that he "was never interested especially in the business end" of aviation to men with no significant connection to Dayton or with no previous aviation experience was not itself misplaced (Curtiss-Wright, the direct successor to the Wright Company, remains in business today), but local hopes for the Wright Company to bring renewed attention to Dayton were. The company's future expansion was not going to occur in the Miami Valley, or even in Ohio.[14]

Wright may not have been interested in the business side of aviation, but the sale of his aviation business left him in comfortable circumstances, allowing him the luxury of never needing to work for a living for the rest of his life. While it took a few months for an accurate price for the Wright Company to appear in public, Wright wound up accepting significantly less money for the firm than the $737,500 he had asked of Morehouse the previous year. The *New York Times* first estimated that Wright received $1.5 million for the company (roughly $33.6 million in 2010), but it later revised the figure to $500,000 (about $11.2 million in 2010) in light of a lawsuit Williamson filed to try to force Wright to pay his fee for brokering the sale. Williamson found that any fraternal connections Wright might have had with a former Dayton neighbor and college classmate of Katharine's soon cooled. Exchanged letters were first addressed to "My dear Orville" or to "Dear Pliny," but soon Wright's letters began with a formal "Dear Sir"; and Williamson claimed he was due a commission of $50,000, 10 percent of the selling price, for acting as broker, while Wright asserted that he was due no more than $25,000. Williamson even pled his case to Katharine and to Lorin Wright, asking them to intercede on his behalf with their brother, and in late October suggested that he and Orville submit their disagreement to independent arbitration. Lorin, acting as his brother's representative, agreed that arbitration might be appropriate, and corresponded with Williamson over proposed arbitration terms. However, disagreements over the principal question under discussion prevented a hearing from taking place. Williamson wanted arbitrators to decide how much he was due and when he should be paid, while Lorin Wright asserted that the judges ought to decide whether his brother was legally liable for a $50,000 payment, or whether any payment to Williamson was voluntary, with the lawyer having "volunteered [his] services to Wilbur and Orville in the personal affairs in connection with the company in consideration of their placing [Williamson] on the directorate of the company," in 1910. Williamson should have kept better track of his old correspondence. He had written the brothers in November 1909 in pursuit of being their "personal representative on that Board of Directors," or in some other corporate capacity, that "the opportunity with its resultant benefits of being brought in contact with your colleagues would be sufficient remuneration for any service I could render to you." Orville Wright used Williamson's words against him. Unable to agree on the main question to place before a panel of arbitrators, the dispute moved to the U.S. District Court for the Southern District of New York. There, Williamson was unable to provide sufficient documentation to prove that Wright had agreed to a 10 percent commission, and his rather weak lawsuit backfired to the extent that he eventually settled the case with Wright for $20,000—$5,000 less than what Wright had initially offered, and $30,000 less than Williamson first claimed as his due.[15] The dispute ended Williamson's relationship with Wright. While the Scarsdale-based lawyer went on to have a successful political career and Wright remained a prominent father of aviation, the two sons of Dayton made no efforts to reconnect.[16]

Meanwhile, the Wright Company was under new ownership. At first, little changed. Manville, Thompson, and Wiggin, though they stayed in New York, did not immediately uproot the airplane factory from its Dayton home. Instead, they brought in new design talent for the production of Models K and the L, though neither was commercially successful. They also took their time assuming some of the company's financial obligations, with Wright, rather oddly, continuing to front the company's weekly payroll of approximately $315 as late as February 1916 from his personal checking account. He was not reimbursed. He and Katharine also journeyed to New York at the end of April for a meeting of Wright Company engineers. But aside from helping with the payroll for a few months, Wright's tenure as a corporate executive was over. With the company sold, Orville retired to the privacy of Hawthorn Hill, to summers on his Lambert Island retreat in Ontario's Georgian Bay, and to experimenting in a laboratory and machine shop he built half a block west of the famous bicycle shop where he and Wilbur built their first airplanes and had kept their office while presiding over the Wright Company. Wright relinquished the lease on that building and all but relinquished an active role in aeronautical research and development. For the remaining thirty-three years of his life, he was feted by politicians, pilots, and other prominent people as an elder statesman of aviation. He served as a consultant to his friend Charles Kettering in the development of the Kettering Bug, an unmanned flying torpedo, in 1918. With James Jacobs, a former Wright Company employee, Wright coinvented a split flap for use on airplanes, for which he and Jacobs received a patent, Wright's last, in 1924. Presidential recognition came in 1920, when Woodrow Wilson nominated Wright to serve on the board of the National Advisory Committee for Aeronautics, a federal agency, founded in 1915, that promoted aviation and aeronautical research and featured the Wrights' 1903 first flight on its official seal. Wright sat on the NACA's main committee until his death and reliably attended board meetings, but he had little active involvement in or practical influence on aeronautical research and development for the rest of his life. Wright, according to Alex Roland, a historian of the NACA, "was on the Committee to grace the letterhead and to add the weight of his reputation to the NACA name," not to lead policy discussions. With Wilbur dead, Orville was content to be a doting uncle, especially to his brother Lorin's children, and avoided the public limelight.[17]

Wright was a Dayton resident for the remainder of his years. His company, though, was not. With no personal ties to southwestern Ohio, Manville and the other new directors believed that Dayton did not have the allied industries needed to support a profitable airplane maker and was not the right location for the company's factory. Manville wanted the company to look beyond its own resources and offer products with major components—especially engines—built elsewhere, especially by companies that specialized in building engines. Dayton's industrial base, the new owners felt, did not provide sufficient options in that field. A few years earlier, Manville might have found options, perhaps from the Speedwell Motor Car Company, which built engines for its cars. But the 1913 flood had not been kind to Pierce Schenck's business, which entered receivership in 1915. Stoddard-Dayton, another prominent local automobile manufacturer, failed in 1913. Though General Motors would later produce cars and car parts in factories around Dayton, those factories lay in the future; Dayton in 1916 was not home to an existing major builder of engines. So, to obtain a "high class reliable aeronautic motor" for its products, the new owners combined the Wright Company with the Simplex Automobile Company of New Brunswick, New Jersey, little more than a month after taking over. Simplex president Henry Lockhart, Jr., became Wright Company president. After being led by the inventors of the airplane, the firm was now in the hands of a banker with extensive business experience. His car company built approximately five hundred luxury automobiles a year and had the facilities in which the Wright Company could build licensed Hispano-Suiza engines, common in Allied airplanes in

Europe, for the company's airplanes. Lockhart also planned to increase the company's production, expanding the Simplex factory in New Brunswick and the Electric Boat Company plant in Bayonne, New Jersey (a company of which new director Henry R. Sutphen was also an officer) to provide the facilities a growing airframe builder would need.[18]

The Wright Company bustled about New Brunswick, but the company grew increasingly quiet in the Gem City. By August 1916, when the Wright Company merged with the Glenn L. Martin Company of Los Angeles, the Dayton factory was merely "an experiment station," one where there were "now not more than 20 receiving regular employment." Huffman Prairie, where the Wrights first flew in 1904, became an afterthought as well. Management closed the company's aviation school there, instead offering budding pilots training at Long Island's Hempstead Plains or at Augusta, Georgia. Even those who still worked at the factory off West Third Street were soon looking for employment elsewhere, though in Dayton's booming industrial economy they likely did not search for long, for by February 1917 the Wright-Martin Company had closed the plant and moved its operations entirely to New Jersey. The closure went unnoticed in the local community, with no reaction by the local papers or by Milton Wright. The dream of Dayton being the home of the American aviation industry was over.[19]

But the Wright name in corporate aviation lived on, and even regained its sole status, without a hyphen, for a time. Glenn Martin, feeling that he had little influence in determining the combined company's direction, left Wright-Martin and formed a new Glenn L. Martin Company in Cleveland in September 1917. Cognizant that it made little business sense to include a competitor's name in its own name, and looking to put "the operations of the older corporation on a satisfactory peace-time basis," Wright-Martin became the Wright Aeronautical Corporation in October 1919. Never able to regain the prominence in airframes that the Wright Company once held with the Model B, Wright Aeronautical shifted its production toward engines and ceased making airplanes in 1926. Curtiss Aeroplane, though, continued to build airplanes. In one of the more ironic corporate mergers of the twentieth century, given the icy relationship of their founders, Wright Aeronautical merged with the Curtiss Aeroplane and Motor Company in 1929 to become the Curtiss-Wright Corporation. Both Curtiss and Wright were so far removed from their respective companies that an uninformed officer of the new firm told reporters that Hammondsport was "the place where the Curtiss brothers came from." Curtiss-Wright remains a maker of aircraft components in the early twenty-first century. Though spread across the globe in 2013, Curtiss-Wright Controls Defense Solutions still maintained an office in Fairborn, Ohio, near Wright-Patterson Air Force Base (which itself surrounds the Wrights' Huffman Prairie proving ground), a vestigial corporate tie to the Dayton region and the birthplace of one of its component ancestors.[20]

# EPILOGUE

## The Wright Company's Legacy

Physically, the Wright Company left a slight legacy. Few of the airplanes its workers built remained intact, and its archives are dispersed and incomplete. Grover Bergdoll's Model B, at the Franklin Institute, and a skeletal Model G exhibited by Dayton Aviation Heritage National Historical Park are two of the most accessible Wright Company airplanes exhibited in museums. The factory still stands, engulfed by later industrial development, but until the National Park Service took an interest in the factory site, only a few plaques, inaccessible behind a secured gate, served to physically commemorate the firm. Its intangible legacy, though, is more significant. As the brothers' last significant collaboration, the fate of the Wright Company demonstrates the extent to which Wilbur and Orville relied on each other for success. The Wright Company's best years, 1910 and 1911, came before Wilbur Wright's untimely death the next year. Without him, the company began its slide to irrelevancy. Left without his brother's leadership and counsel, Orville was unprepared and uncomfortable as the president of a well-capitalized corporation that had as its directors titans of Wall Street with national aspirations. He trusted no one else to the extent that he trusted Wilbur, and he neither solicited management advice from the company's broadly experienced board members or employees nor welcomed it when it arrived on its own accord, as in the controversy over whether to hire William McCombs as a company lobbyist. How a Wright Company in which Wilbur Wright still played a role would have responded to the army's problems with the Model C, the shift from pusher to tractor airplanes, or the 1914 court victory over Curtiss for patent infringement will never be known, but a company led by two brothers who used their bickering yet symbiotic relationship to invent the airplane would have developed differently than did the Wright Company under Orville Wright's sole direction.

The Wrights' litigiousness produced its own results besides the court verdicts. The Manufacturers Aircraft Association, which continued to cross-license aviation patents until its dissolution, in 1975, was a direct legacy of the Wright Company's patent battles with Glenn Curtiss's firms and of the U.S.

government's frustrations with the effects of the court cases on the industry's domestic development. As late as 1972, in the midst of a federal antitrust lawsuit, the chairman of McDonnell Douglas claimed that the MAA had "made it possible for small companies and new enterprises" to gain footholds in the aviation industry, preventing the development of the monopolized industry that some of the Wright Company's backers wanted to create. And though it never caused technological development to the extent provoked by the arms race in Europe on the eve of the First World War, the Wrights' dogged protection of their patent ironically helped spur the development of alternative technologies, as Curtiss and other builders looked for ways to build airplanes that would not provoke lawsuits from Dayton. Curtiss doggedly defended the aileron as being a means of controlling an airplane not covered by the 1906 Wright patent, and U.S. companies generally followed his lead. Curtiss and his company held key patents on the aileron, and the patent pool run by the MAA ensured that other manufacturers (including the Wright Company) could use these patents without fear of a lawsuit and spread the adoption of his developments through the rest of the industry.[1]

Indeed, the Wright Company's litigiousness preceding the establishment of the Manufacturers Aircraft Association greatly affected early aviation history in the United States. The varying reactions of other aviators and airplane builders to Judge John R. Hazel's 1913 ruling show just how much perceived power and influence the Wright Company retained even as few of its airplanes remained in North American skies. To some, it was a giant that should not be disturbed. The March 1913 issue of *Aeronautics* included several articles on the verdict and editorialized that "everyone will be relieved at last to see the patent adjudicated" and that it was "doubtless true that capital has been 'scared off' due to the uncertainty." With the legal status of the patent resolved in favor of the Wright Company, *Aeronautics* editors believed that inventors would turn to developing noninfringing stabilizers or would not "object to paying a moderate royalty." The adjudicated patent remained one of the Wright Company's most valuable assets throughout its existence and was one of the few lures available to Orville Wright as he tried to sell the company to new ownership. The potential of a new airplane design to conflict with the Wright patent remained a point for discussion until the MAA's creation.[2]

OTHER companies soon engulfed the physical legacy of the Wright Company in Dayton. During their century of industrial use, the company's factory buildings would always be connected with transportation. In March 1917, Wright-Martin sold the property to the Darling Motor Company, an automobile maker formed by several NCR employees for $40,000. Unfortunately for Darling, the market for its cars crashed after the United States entered the First World War the next month, and it quickly went out of business, leaving the buildings again vacant just as the United States needed to produce more and more war matériel. The U.S. aviation industry was still more underdeveloped than its European siblings, and the army suddenly needed more airplanes in its arsenals as its soldiers crossed the Atlantic. Some of Dayton's business leaders believed that the vacant buildings could be used again for airplane construction.[3]

Five days after the United States entered the war, local industrialists Edward Deeds, Charles Kettering, Harold E. Talbott, Sr., and Harold E. Talbott, Jr., looking to capitalize on the army's need for airplanes and on Dayton's industrial and aviation histories, organized the Dayton Airplane Company, which they soon rebranded as the Dayton-Wright Airplane Company, with the elder Talbott, an engineer, as chairman and his son as president.[4] Orville Wright served the firm as a consulting engineer, working on the company's contributions to the Liberty engine and the Kettering Bug unmanned flying torpedo, but was not involved in its management. With vacant factory space in the Dayton area difficult to find, Dayton-Wright purchased the former Wright Company buildings from the bankrupt Darling Motor Company in early 1918. Though most of the company's assembly work occurred in Moraine, workers made metal fittings and undertook specialized priority projects in the Wright Company's former buildings, now Dayton-Wright's Plant 3. Dayton-Wright was one of the largest U.S. builders of British-designed de Havilland-4 bombers, the only airplane produced in the United States that was used in Europe by Allied air forces. Even as women took up new jobs, its production floor remained heavily male, as the company benefited from the Aircraft Production Board securing draft exemptions for all workers in the United States involved in building airplanes for the military (though nearly 24 percent of aircraft factory workers industrywide would be female at the war's end). While that was a significant percentage, the phenomenon of Rosie the Riveter and her spread throughout factories would have to wait another twenty years for a war in which the United States fought for a longer time.[5]

Meanwhile, the U.S. government, which wanted airplane builders to make airplanes instead of legal briefs, acted to end one of the most serious brakes on domestic aviation: the industry's divisive patent situation. Through Secretary of War Newton Baker, Secretary of the Navy Josephus Daniels, and the National Advisory Committee for Aeronautics, the Wilson administration pressured the industry in 1917 to create a patent pool for aircraft licensing managed by the Manufacturers Aircraft Association. The new organization, of which Frank Russell became president, instituted a $200 royalty on each aircraft made by a member company, initially sending $135 of the fee to Wright-Martin, $40 to Curtiss, and $25 to the organization's treasury. But even with the end of the patent war, European-designed and European-built models dominated the skies of Europe. U.S. workers made few airplanes based on domestic patents during the war, and none flew over the trenches of the Western Front.[6]

The Dayton-Wright Airplane Company survived postwar demilitarization, but not on its own. Partly to obtain the research and development skills of Charles Kettering, who, among his half-dozen corporate offices, was a Dayton-Wright vice president and consulting engineer, General Motors acquired Dayton-Wright (then worth $1.2 million) for 10,960 shares of its own stock in September 1919. GM reincorporated the airplane maker a few months later as the Dayton Airplane Company but soon decided that building airplanes did not fit with its other lines of business. In 1923, GM left the airplane industry and turned the former Wright Company buildings into the core of a factory for its Inland Manufacturing Corporation subsidiary. Throughout the rest of the twentieth century, Inland, and later Delco and Delphi, workers made wood veneer–wrapped iron steering wheels, brake linings, and other automobile parts at the site, until Delphi, as part of Chapter 11 bankruptcy proceedings, closed its factory in 2008. A year later, the U.S. Congress added the Wright Company's 1910 and 1911 buildings and three later structures to the boundary of the National Park Service's Dayton Aviation Heritage National Historical Park.[7]

Led by brothers with little experience in running a business with a national scope, two men who were talented inventors and engineers but who resisted the suggestions of their company's managers and stockholders to create corporate growth, the Wright Company never became the dominant builder of airplanes its creators had hoped it would become in 1909. Instead, Wilbur and Orville Wright saw that it devoted much of its corporate attention to protecting the 1906 Wright patent from infringement at the expense of product development, resulting in the company being an insignificant part of the industry when Orville Wright sold it, in 1915. The Wright Company adapted too slowly to technological developments for its airplanes to be a significant part of the U.S. civilian or military fleets after 1912. Even though he lost the

FIGURE E.I. Pliny W. Williamson, Katharine Wright, Milton Wright, John R. McMahon, Horace Wright (*seated, left to right*) and Orville Wright and Earl N. Findley (*standing, left to right*) on the east porch of Hawthorn Hill, Oakwood, Ohio, 1915. Findley and McMahon wrote about aviation. *Courtesy of the Library of Congress*

patent-infringement lawsuits filed by the Wright Company, Glenn Curtiss expanded his operations, evolving from a small operation in rural Steuben County, New York, to one with multiple factories employing hundreds in Buffalo and Toronto, factories that began to incorporate aspects of assembly line manufacturing as the United States approached its decision to enter the First World War. Meanwhile, Starling Burgess's company proved successful enough that in 1916 Curtiss purchased the Massachusetts firm, which, ironically, Frank Russell had managed since late 1911. In Dayton, the Wright Company remained a smaller operation than its northeastern competitors, staffed at its height by at most sixty skilled craftspeople, a company blending a late-nineteenth-century labor model with a twentieth-century product.

While not a successful airplane builder, the Wright Company is still recognized as a significant early aeronautical concern. It was the lodestar around which the rest of the industry moved, trying to fight or avoid its patent infringement lawsuits. Its namesakes remain the most prominent of early U.S. aviators. Its litigiousness prompted the creation of the Manufacturers Aircraft Association. Moreover, the company provided early industry experience to future industry executives Frank Russell and Grover Loening, entertained thousands of people through its exhibition department, and trained several hundred individuals to fly through its aviation schools. The eventual irrelevance of the Wright Company demonstrates the importance of combining advanced engineering and new technologies with competent business management to ensure a company's success. Even though it became an industry afterthought, the prominence of its presidents made the company significant in the transformation of the airplane from a curious wonder into a serious means of transportation. Though the aeronautical industry quickly grew to cities throughout North America and names other than Wright gained fame through aviation, by the time Orville sold the Wright Company, he had truly succeeded in his and his brother's dream of making airplanes practical vehicles with multiple uses.

# NOTES

## Abbreviations

Knabenshue Papers    A. Roy Knabenshue Papers, Archives Division, National Air and Space Museum, Smithsonian Institution

Loening Papers    Grover C. Loening Papers, Library of Congress

NASM    National Air and Space Museum, Smithsonian Institution

Russell diary    Frank Henry Russell Papers, diary, American Heritage Center, University of Wyoming, Laramie

WBP    Wilbur and Orville Wright Papers, Manuscript Division, Library of Congress

WCP    Wright Company Papers, Museum of Flight, Seattle

## Preface

1. In early 2013, only six Wikipedias—the Chinese, English, Finnish, German, Italian, and Spanish editions—contained articles, each only a few sentences, about the Wright Company.

2. Tom D. Crouch, *The Bishop's Boys: A Life of Wilbur and Orville Wright* (New York: Norton, 1989); Fred Howard, *Wilbur and Orville: A Biography of the Wright Brothers* (New York: Knopf, 1987); Julie Hedgepeth Williams, *Wings of Opportunity: The Wright Brothers in Montgomery, Alabama, 1910* (Montgomery: NewSouth Books, 2010); John Carver Edwards, *Orville's Aviators: Outstanding Alumni of the Wright Flying School, 1910–1916* (Jefferson, NC: McFarland, 2009); John S. Olszowka, "From Shop Floor to Flight: Work and Labor in the Aircraft Industry, 1908–1945" (PhD diss., Binghamton University, 2000); C. R. Roseberry, *Glenn Curtiss: Pioneer of Flight* (1972; repr., Syracuse, NY: Syracuse University Press, 1991); Seth Shulman, *Unlocking the Sky: Glenn Hammond Curtiss and the Race to Invent the Airplane* (New York: HarperCollins, 2002); Bartlett Gould, "Burgess of Marblehead," *Essex Institute Historical Collections* 106, no. 1 (January 1970): 3–31; Donald M. Pattillo, *Pushing the Envelope: The American Aircraft Industry* (Ann Arbor: University of Michigan Press, 1998); Wayne Biddle, *Barons of the Sky: From Early Flight to Strategic Warfare; The Story of the American Aerospace Industry* (New York: Henry Holt, 1991); Judith Sealander, *Grand Plans: Business Progressivism and Social Change in Ohio's Miami Valley, 1890–1929* (Lexington: University Press of Kentucky, 1988); John T. Walker, "Socialism in Dayton, Ohio, 1912 to 1925: Its Membership, Organization, and Demise," *Labor History* 26, no. 3 (Summer 1985): 384–404; Richard W. Judd, *Socialist Cities: Municipal Politics and the Grass Roots of American Socialism* (Albany: State University of New York Press, 1989); Lindy Biggs, *The Rational Factory: Architecture, Technology, and Work in America's Age of Mass Production* (Baltimore: Johns Hopkins University Press, 1996); Daniel Nelson, *Farm and Factory: Workers in the Midwest, 1880–1990* (Bloomington: Indiana University Press, 1995); David A. Hounshell,

*From the American System to Mass Production, 1800–1932: The Development of Manufacturing Technology in the United States* (Baltimore: Johns Hopkins University Press, 1984).

3. Douglas Gantenbein, "Aviation's Birth Certificate," *Air and Space Magazine* 17, no. 6 (March 2003): 78–79.

4. Richard P. Hallion, "The Wright Kites, Gliders, and Airplanes: A Reference Guide," 19 August 2003, www.af.mil/shared/media/document/AFD-051013-002.pdf; Wilbur Wright, Orville Wright, and Octave Chanute, *The Papers of Wilbur and Orville Wright, Including the Chanute-Wright Letters and Other Papers of Octave Chanute*, ed. Marvin W. McFarland (1953; repr., Salem, NH: Ayer, 1990), 2:1183–1210; Paul Glenshaw, "Ladies and Gentlemen: The Aeroplane!" *Air and Space Magazine* 23, no. 1 (April–May 2008): 48–55; Glenshaw, "Kings of the Air," *Air and Space Magazine* 27, no. 7 (February–March 2013): 46–51; Edwards, *Orville's Aviators*; Williams, *Wings of Opportunity*.

## Chapter 1: "We Will Devote . . . Our Time to Experimental Work"

1. Orville Wright and Wilbur Wright, Flying-machine, U.S. Patent 821,393, filed 23 March 1903, and issued 22 May 1906. For the GDP figure, see Stephen N. Broadberry and Douglas A. Irwin, "Labour Productivity in the United States and the United Kingdom during the Nineteenth Century" (discussion paper, Centre for Economic Policy Research, London, 2005), 22.

2. "Notes of Conversation between Wilbur and Orville Wright and Hart O. Berg, Paris, 6 November 1907," in Wilbur Wright, Orville Wright, and Octave Chanute, *The Papers of Wilbur and Orville Wright, Including the Chanute-Wright Letters and Other Papers of Octave Chanute*, ed. Marvin W. McFarland (New York: McGraw-Hill, 1953; repr., Salem, NH: Ayer, 1990), 2:832.

3. Tom D. Crouch, *The Bishop's Boys: A Life of Wilbur and Orville Wright* (New York: Norton, 1989), 371–90.

4. Ibid., 33. For an extensive examination of the brothers' printing business, see Charlotte K. Brunsman and August E. Brunsman, *The Other Career of Wilbur and Orville: Wright & Wright Printers* (Kettering, OH: Trailside Press, 1989).

5. "Eighty Aeroplanes Ordered in America," *New York Times*, 24 May 1909, 3.

6. "Dayton's Seventh Flood," *Boston Evening Transcript*, 27 March 1913, 3; Campbell Gibson, "Population of the 100 Largest Cities and Other Urban Places in the United States: 1790 to 1990," Population Division Working Paper 27 (Washington, DC: U.S. Bureau of the Census, Population Division, 1998), www.census.gov/population/www/documentation/twps0027/twps0027.html; Gibson, "Population of the 100 Largest Urban Places: 1900," www.census.gov/population/www/documentation/twps0027/tab13.txt; U.S. Department of Commerce and Labor, Bureau of Foreign and Domestic Commerce, *Statistical Abstract of the United States 1912, Thirty-fifth Number* (Washington, DC: Government Printing Office, 1913), 49, 70; John T. Walker, "Socialism in Dayton, Ohio, 1912 to 1925: Its Membership, Organization, and Demise," *Labor History* 26, no. 3 (Summer 1985): 389n11; Peter S. Cajka, "The National Cash Register Company and the Neighborhoods: New Perspectives on Relief in the Dayton Flood of 1913," *Ohio History* 118 (2011): 51.

7. Carl Becker, "A 'Most Complete' Factory: The Barney Car Works, 1850–1926," *Cincinnati Historical Society Bulletin* 31, no. 1 (Spring 1973): 66.

8. Walker, "Socialism in Dayton," 386; Fred Mitchell, "Historic and Architectural Resources of the Webster Station Area, Dayton, Ohio," National Register of Historic Places Multiple Property Documentation Form (2000), E3; Carl Becker, "Mill, Shop and Factory: The Industrial Life of Dayton, Ohio, 1830–1900" (PhD diss., University of Cincinnati, 1971), 340; Judith Sealander, *Grand Plans: Business Progressivism and Social Change in Ohio's Miami Valley, 1890–1929* (Lexington: University Press of Kentucky, 1988), 19–25.

9. U.S. Department of Commerce, Bureau of the Census, *Thirteenth Census of the United States Taken in the Year 1910*, vol. 3, *Population—Reports by States, Nebraska-Wyoming, Alaska, Hawaii and Porto Rico* (Washington, DC: Government Printing Office, 1913), 429; U.S. Department of Commerce, Economics and Statistics Administration, Bureau of the Census, *Statistical Abstract of the United States: 2003* (Washington, DC: Government Printing Office, 2003), 37, www.census.gov/statab/hist/HS-22.pdf; Augustus W. Drury, *History of the City of Dayton and Montgomery County, Ohio* (Chicago: S. J. Clarke, 1909), 1:452–53, 463–64, 466, 468–69. Bonebrake Theological Seminary became United Theological Seminary in 1954.

10. U.S. Department of Commerce, Bureau of the Census, *Thirteenth Census of the United States Taken in the Year 1910*, vol. 4, *Population: Occupational Statistics* (Washington, DC: Government Printing Office, 1913), 152–65.

11. James M. Cox to John N. Wheeler, 19 April 1939, box 6, file 13, James M. Cox Papers, University Library, Wright State University; Milton Wright, *Diaries, 1857–1917* (Dayton: Wright State University, 1999), 634, 652; Sealander, *Grand Plans*, 85–128.

## Chapter 2: Bringing an Aeroplane Factory to Dayton

1. "Peterkin Not Around," *New York Sun*, 9 December 1909, 5; "Obituary Notes," *New York Times*, 19 November 1895, 5; "Mrs. Helena Peterkin, Charity Worker, Dies," *New York Times*, 3 April 1926, 17; Clinton R. Peterkin to Orville Wright, 18 December 1931, box 52, Wilbur and Orville Wright Papers, Manuscript Division, Library of Congress (hereafter WBP); Arnold Kruckman, "Obscure Young Peterkin and His Aeroplane Trust," *New York World*, 28 November 1909, 4; Tom D. Crouch, *The Bishop's Boys: A Life of Wilbur and Orville Wright* (New York: Norton, 1989), 409; Fred C. Kelly, *The Wright Brothers*, 2nd ed. (New York: Farrar, Straus and Young, 1950), 269.

2. Wilbur Wright to Octave Chanute, 6 December 1909, box 42, Special Correspondence—Wright Brothers, 1908–1909, Octave Chanute Papers, Manuscript Division, Library of Congress.

3. Interborough Rapid Transit Company, *Interborough Rapid Transit: The New York Subway; Its Construction and Equipment* (New York: author, 1904; repr., New York: Arno, 1969), 10; "Gossip of Wall Street," *New York Sun*, 24 November 1909, 11.

4. "The Wright Company," corporate disclosure and stock-offering prospectus, 9 November 1909, box 1, folder 9, Wright Company Papers, Museum of Flight, Seattle (hereafter WCP); list of April 1914 Stockholders, General Correspondence: Williamson, Pliny W., 1914, box 65, WBP; "Minutes of Board of Directors Meeting, 6 January 1911, WCP; "Alger a Director in Wright Company," *Detroit Free Press*, 10 December 1909, 14; "The Wrights Tie Up Air Meet Receipts," *New York Times*, 8 December 1910, 4; "Allan Ryan's Career in Wall St. Stormy," *New York Times*, 22 July 1922, 3; "Ryan Forms Another

Aeroplane Company," *New York World*, 23 October 1911, morning edition; "Court Circular," *Times* [London, England], 24 November 1909, 13; Robert J. Collier to Orville Wright, n.d., box 20, General Correspondence: Collier, Robert J., 1910–18, WBP. Frank S. Hedley, general superintendent of the IRT, also owned Wright Company stock, submitting proxies from 1911 to 1913, though he is never listed as a director. Howard Gould (1871–1959) was a financier and the son of railroad developer Jay Gould. Ryan, whose relationship with his wealthy father collapsed over Thomas F. Ryan's remarriage less than two weeks after his first wife's death in 1917, went on to a prominent Wall Street career in which he gained notoriety for declaring bankruptcy after cornering the stock of the Stutz Motor Company, a luxury automobile builder, in 1920. See "Allan Ryan Dies; 'Cornered' Stutz," *New York Times*, 27 November 1940, 29.

5. Wright Company, Articles of Incorporation, Subject File: Wright Company—Establishment of Corporation, 1909–10, n.d., box 108, WBP; "Night and Day Bank Building," *United States Investor* 20, no. 43 (6 November 1909): 1986. The company's executive committee minutes for 11 January 1910 suggest that the lease of suite 602 in the Night and Day Bank Building expired on 1 May 1912; the New York office may have shifted to 11 Pine Street after that date with the completion of the Bankers Trust Building that year. Esther Beattie, of Orange, New Jersey, worked as a secretary in the New York office in 1910; see "Death of Johnstone Foretold in a Dream," *Washington Times*, 19 November 1910, 13.

6. "Rich Men Take Up the Dangerous Sport of Flying," *New York Times*, 26 June 1910, SM5; "Alger Studies Aviation," *New York Times*, 30 July 1910, 4; Russell Alger to Wilbur Wright, 16 October, 11 November, 2 December 1909, box 10, WBP; Frederick Alger to Wilbur Wright, 13 September 1909, box 10, WBP.

7. Automatic Hook and Eye was not profitable during Russell's tenure as president; it owed him $8,000 in unpaid salary when he went to work for the Wright Company. Russell even tried to drive some Wright Company business to Hoboken, suggesting that Automatic Hook and Eye's machine shop produce small parts for the Dayton plant. See Robert Friedel, *Zipper: An Exploration in Novelty* (New York: Norton, 1994), 75; Russell to Wilbur Wright, 3 January 1910, General Correspondence: Russell, Frank H., 1909–12, 1927–38, 1947, box 55, WBP.

8. "Who's Who in Aeronautics," *Aerial Age Weekly* 13, no. 3 (28 March 1921): 61; entries for 23 July, 27 September, 8, 9, 30 December 1909, 17 January 1910, diary, box 3, Frank Henry Russell Papers, collection 11624, American Heritage Center, University of Wyoming (hereafter Russell diary); Russell Alger to Wilbur and Orville Wright, 29 November 1909, General Correspondence: Alger, Russell A., 1908–9, box 10, WBP; [Wilbur Wright] to Russell, 4 January 1910, General Correspondence: Russell, Frank H., 1909–12, 1927–38, 1947, box 55, WBP.

9. "Big Dayton Plant for Wright Airships," *New York Times*, 25 November 1909, 7; "Hard to Train Men to Build Aeroplanes," *New York Times*, 12 April 1910, 6; Irwin Ellis, "Aviation as a Business: An Infant Industry That Promises to Rival Auto Trade," *Chicago Daily Tribune*, 20 August 1911, E1; "Largest Aeroplane Co. in the World Will Be Represented in Dayton," *Dayton Daily News*, 17 August 1916.

10. Philip Scranton, "Diversity in Diversity: Flexible Production and American Industrialization, 1880–1930," *Business History Review* 65, no. 1 (Spring 1991): 39, 41.

11. City directory entries for employees of such major Dayton employers as NCR, Barney and Smith, and Dayton Malleable Iron generally noted a person's place of employment in addition to the person's job. No records concerning the mechanics of the Wright Company's process for hiring factory workers—or, indeed, of who actually made hiring decisions—remain.

12. David A. Gerber, *Black Ohio and the Color Line, 1860–1915* (Urbana: University of Illinois Press, 1976), 301. The aircraft industry in general was heavily white; see John S. Olszowka, "From Shop Floor to Flight: Work and Labor in the Aircraft Industry, 1908–1945" (PhD diss., Binghamton University, 2000), 139. In 1910, Montgomery County's population was 3.3 percent African American (5,481 individuals).

13. U.S. Department of Commerce and Labor, Bureau of the Census, *Thirteenth Census, 1910, Dayton, Montgomery County, Ohio*, s.v. "Louis Luneke," "Harry Harrold," and "John Steinway," Heritage Quest, www.heritagequestonline.com; Bureau of the Census, *Fourteenth Census, 1920, Youngstown, Mahoning County, Ohio*, s.v. "Frank Quinn," Heritage Quest Online; Bureau of the Census, *Thirteenth Census, 1910, Hammondsport Village, Urbana Township, Steuben County, New York*, s.v. "Lewis Blomquist" and "Elmer E. Robinson," Heritage Quest Online; Massachusetts, Bureau of Statistics, and Charles F. Gettemy, *The Decennial Census, 1915* (Boston: Wright and Potter, 1918), 313–19. In 1914 nearly 71 percent of the workers at the Ford Motor Company's Highland Park plant, in Michigan, were foreign born; see Stephen Meyer, "Adapting the Immigrant to the Line: Americanization in the Ford Factory, 1914–1921," *Journal of Social History* 14, no. 1 (Autumn 1980): 69.

14. Judith Sealander, *Grand Plans: Business Progressivism and Social Change in Ohio's Miami Valley, 1890–1929* (Lexington: University Press of Kentucky, 1988), 16; Williams Directory Company, *Williams' Dayton Directory for 1911–1912* (Cincinnati: Williams Directory Co., 1911), s.v. "Augusta E. Smith," 1181, and "Maude A. Thomas," 1269; entry for 19 January 1910, Russell diary; "Wright Seamstress Makes First Flight," *Hartford Courant*, 22 November 1969, 2; Roberta Cohen, "Former Wright Brothers Employee Tours Delphos Historical Museum," *Delphos Tri-County Daily Herald*, 6 October 1975, 1. For the photographs of company departments, see MS-1, series 22, box 21, file 3, WBP; they are also searchable at http://core.libraries.wright.edu/handle/2374.WSU/827, s.v. "Wright Company."

15. NCR employees worked an eight-hour day, while Barney and Smith employees labored for twelve hours a day, six days a week. The 1912 Ohio Constitution allowed the state legislature to set a minimum wage, but the legislature did not take advantage of that ability.

16. Classified advertisements, *Dayton Daily News*, 24 February, 19, 29 March 1910; Orville Wright to Andrew Freedman, 10 May 1913, General Correspondence: Andrew A. Freedman, box 30, WBP; Tom Russell, interview by Susan Bennet, 24 April 1967, Wright Brothers–Charles F. Kettering Oral History Project, University of Dayton, transcript, Roesch Library, 9; Ernest Dubel, interview by Susan Bennet, 1 March 1967, transcript, Roesch Library, University of Dayton, 4; P. O. Warren, "America's Best Employers," *Forbes*, 16 March 1918; Scott D. Trostel, *The Barney and Smith Car Company: Car Builders, Dayton, Ohio* (Fletcher, OH: Cam-Tech Publishing, 1993), 131, 147.

17. Daniel Nelson, "The New Factory System and the Unions: The National Cash Register Dispute of 1901," *Labor History* 15, no. 2 (1974): 171.

18. See John Kirby, Jr., *The Career of a Family; or, the Ups and Downs of a Lifetime* (Dayton: Otterbein Press, 1916).

19. Crouch, *Bishop's Boys*, 276; Sealander, *Grand Plans*, 30–31; Olszowka, "Shop Floor to Flight," 4, 10; Fred Kreusch, interview by Susan Bennet, 8 March 1967, Wright Brothers–Charles F. Kettering Oral History Project, University of Dayton, transcript, Roesch Library, 6, 7, 17; Horace Wright, "Recollections," in *Wright Reminiscences*, ed. Ivonette Wright Miller (Wright-Patterson Air Force Base, OH: Air Force Museum Foundation, 1978), 158. Crouch also recounts the same mashed-potatoes story. Crouch, *Bishop's Boys*, 127.

### Chapter 3: "A Substantial, Commodious, Thoroughly Modern Factory"

1. "Aeroplane Factory," *Dayton Daily News*, 22 December 1909, 6; "Aeroplane Factory to Come to Dayton," *Dayton Herald*, 29 November 1909, 12; "Wrights Buy 700 Acres," *New York Times*, 7 September 1909, 6; "W. Wright Denies Buying Land at Tippecanoe City," *Dayton Herald*, 6 September 1909, 9; "No Truth in Story of Wright Factory," *Dayton Daily News*, 6 September 1909, 4; "Wright Factory May Be Moved?" *Dayton Daily News*, 21 May 1910, 16; "Dayton's Big Chance Is at Hand to Secure New Wright Aeroplane Plant," *Dayton Herald*, 23 November 1909, 1, 16. The story about the supposed Tippecanoe City land purchase also ran in the *Fredericksburg (VA) Free Lance*, the *Mahoning (OH) Dispatch*, the *Marion (OH) Daily Mirror*, and the *Willmar (MN) Tribune*.

2. Charlotte K. and August E. Brunsman, "Friendship Carved in Stone," photocopy, in possession of author; "Wrights to Build a Plant," *Dayton Journal*, 6 September 1909, 1; Wright Company executive committee minutes for 29 September 1910, WCP; entries for 24, 26, 27, 29 August, 3 September, 26, 29 October 1910, Russell diary; "Large Factory Site Is Purchased by the Wrights," *Dayton Daily News*, 21 August 1910, 1; Drury, *History of the City*, 237, 373–72, 992–93.

3. "Oakwood Man Dies Suddenly Early Thursday," *Dayton Herald*, 11 June 1918, 14; "Wright Brothers Ask for One Thousand Feet of Floor Space," *Dayton Daily News*, 28 November 1909; Wright brothers to Russell Alger, 4 December 1909, box 8, General Correspondence: Alger, Russell A., 1908–9, WBP; "Speedwell Builds Aeroplane Plant," *Fort Worth Star-Telegram*, 25 December 1909, 19; "Deal Is All But Closed for Wright Factory Site," *Dayton Herald*, 27 August 1910, 6.

4. Entries for 7, 22 November 1910, Russell diary; "Wright Aviators Busy with Many Exhibitions," *Dayton Daily News*, 8 October 1910, 10; Irwin Ellis, "Aviation as a Business: An Infant Industry That Promises to Rival Auto Trade," *Chicago Daily Tribune*, 20 August 1911, E1. For a discussion of daylight factories, see Lindy Biggs, *The Rational Factory: Architecture, Technology, and Work in America's Age of Mass Production* (Baltimore: Johns Hopkins University Press, 1996), 96–100.

5. Wright brothers to Russell A. Alger, 29 August 1910, General Correspondence: Alger, Russell A., 1910–11, box 8, WBP; Wright Company executive committee minutes for 29 September 1910, WCP; Wright Company, *Wright Flyers* (New York: Premier Press, 1912), in box 108, WBP; Ellis, "Aviation as a Business"; "Manufacturers Make Product Suggestions," *Aviation and Aeronautical Engineering* 4, no. 11 (1 July 1918): 781.

6. Fred Howard, *Wilbur and Orville: A Biography of the Wright Brothers* (New York: Knopf, 1987), 352; Tom D. Crouch, *The Bishop's Boys: A Life of Wilbur and Orville Wright*

(New York: Norton, 1989), 424; William Conover, interview by Susan Bennet, 4 February 1967, Wright Brothers–Charles F. Kettering Oral History Project, University of Dayton, transcript, Roesch Library, Dayton, 1; [Wilbur?] Wright to Alger, 29 August 1910, General Correspondence: Alger, Russell A., 1910, box 10, WBP; "Speedwell Builds Aeroplane Plant," *Fort Worth Star-Telegram*, 25 December 1909, 19; Herbert A. Johnson, *Wingless Eagle: U.S. Army Aviation through World War I* (Chapel Hill: University of North Carolina Press, 2001), 110–11; "Aeroplane Shipment from Wright Plant," *Atlanta Constitution*, 16 June 1910, 6. For the length of time required to deliver a Wright G, see Orville Wright to A. B. Gaines, Jr., telegram, 19 April 1915, General Correspondence: Gaines, A. B., 1913–14, box 31, WBP; Howard Rinehart to Orville Wright, 17 August 1915, General Correspondence: Rinehart, Howard, 1915–18, box 54, WBP. Both the Speedwell and west Dayton plants had their own railroad sidings, providing the Wright Company quick access to the U.S. railway network.

7. Fred T. Jane, ed., *Jane's All the World's Airships, 1909* (London: Sampson Low, Marston, 1909; repr., New York: Arco, 1969), 92 (French industry, 92–158; U.S. industry, 251–304b).

8. "Building Aeroplanes in Paris: A Thriving Industry That Calls Many New Theories into Play," *New York Times*, 25 April 1909, SM4; Tom D. Crouch, "Blaming Wilbur and Orville: The Wright Patent Suits and the Growth of American Aeronautics," in Peter Galison and Alex Roland, eds., *Atmospheric Flight in the Twentieth Century* (Dordrecht: Kluwer Academic, 2000), 295; Emmanuel Chadeau, *De Blériot à Dassault: Histoire de l'industrie aéronautique en France, 1900–1950* (Paris: A. Fayard, 1987), 37, 61; Warren I I. Miller, "The Manufacture of French Aeroplanes," *Engineering Magazine* 39, no. 5 (August 1910): 649; Carl Dienstbach, "The Rise of the Flying Machine Industry in America," *American Aeronaut* 1, no. 1 (August 1909): 12; Juliette Hennessy, *The United States Army Air Arm, April 1861 to April 1917* (1958; repr., Washington, DC: Office of Air Force History, U.S. Air Force, 1985), 134. References to one of Henri Farman's early airplanes being stored on the factory floor suggest that the *Times* reporter may have visited the Farman Aviation Works.

9. "Wrights to Fight Foreign Airplanes," *New York Times*, 25 September 1909, 1; "Blériot Machines Coming," *New York Times*, 30 September 1910, 5; Ernest Jones, "A Review of 1911—Forecast for 1912," *Aeronautics* 10, no. 1 (January 1912): 1; "The Motor," *Evening Post* (Wellington, New Zealand) 83, no. 46 (23 February 1912): 10; Tom D. Crouch, *Wings: A History of Aviation from Kites to the Space Age* (New York: Norton, 2003), 134–35, 147; "New Companies," *Aeronautics* 10, no. 5 (May–June 1912): 178–79. Jones noted that Curtiss built approximately twenty-five airplanes in 1911; production numbers for the Wright Company that year are not available.

10. C. R. Roseberry, *Glenn Curtiss: Pioneer of Flight* (1972; repr., Syracuse, NY: Syracuse University Press, 1991), 32, 34, 158, 164, 257.

11. Augustus Post, "A Visit to the Factory," in *The Curtiss Aviation Book*, by Glenn Hammond Curtiss and Augustus Post (New York: Frederick A. Stokes Co., 1912), 304–6; "Curtiss to Move His Plant," *New York Times*, 6 December 1914, 1. For a recollection by Wright Company employee Frederick J. Kreusch of experimenting on engines at the former Wright Cycle Company building, see "Co-Inventors Made Good Team, Recalls Bicycle Shop Worker," *Dayton Journal Herald*, 17 December 1953.

12. W. C. Beers's U.S. Aeronautical Company of New Haven, Connecticut, obtained a manufacturing license from the Wright Company in May 1911, but it never commercially

produced Wright-type airplanes. See Alpheus Barnes to Orville Wright, 8 May 1911, WCP. Extant Wright Company papers do not contain a copy of the Burgess license.

13. "W. S. Burgess Dies; Yacht Builder, 68," *New York Times*, 20 March 1947, 27; Harvard College, Class of 1901, *Secretary's Third Report* (Cambridge, MA: Crimson Printing Office, 1911), 68; Alpheus F. Barnes to Frederick P. Fish, 22 June 1914, General Correspondence: Fish, Richardson and Neave, 1914, box 29, WBP; "Millionaire Society Men Learning to Fly in Wright Aviation School at Augusta," *Atlanta Constitution*, 12 February 1911, 11; John Carver Edwards, *Orville's Aviators: Outstanding Alumni of the Wright Flying School, 1910–1916* (Jefferson, NC: McFarland, 2009), 131. For an airplane-focused history of Burgess's aviation career, see Bartlett Gould, "Burgess of Marblehead," *Essex Institute Historical Collections* 106, no. 1 (January 1970): 3–31. Note the difference in spelling between Greely Curtis and Glenn Curtiss; they were not related.

14. Bowd Osborne, photographs and color slides, Marblehead Historical Commission, http://www.marbleheadhistory.com/exhibit2/vexmain2.htm, "Search this exhibit," s.v. "Burgess"; Greely S. Curtis to Wilbur Wright, 9 November 1911, General Correspondence: Burgess Company and Curtis, 1910–11, box 16, WBP; "The Curtiss Aeroplane and Motor Company Acquires Services of W. Starling Burgess and Burgess Company," *Aerial Age Weekly* 2, no. 22 (14 February 1916): 519; "Curtiss Gets Burgess Co.," *New York Times*, 10 February 1916, 15; "The Burgess Company Expanding," *Aerial Age Weekly* 3, no. 2 (27 March 1916): 58.

15. David A. Hounshell, *From the American System to Mass Production, 1800–1932: The Development of Manufacturing Technology in the United States* (Baltimore: Johns Hopkins University Press, 1984), table 6.1, 224; Donald M. Pattillo, *Pushing the Envelope: The American Aircraft Industry* (Ann Arbor: University of Michigan Press, 1998), 29; Dubel, interview by Susan Bennet, 1 March 1967, transcript, 3. Of the 305 airplanes ordered but not delivered between 1907 and 1917, orders for 155 were cancelled while 150 were delivered after the United States entered the First World War. See James C. Fahey, *U.S. Army Aircraft (Heavier-than-Air) 1908–1946* (Falls Church, VA: Ships and Aircraft, 1946), 6.

## Chapter 4: "Our Machines Are Sold on Their Merits"

1. Wilbur Wright to Hart Berg, 16 October 1910, in Wilbur Wright and Orville Wright, *Miracle at Kitty Hawk: The Letters of Wilbur and Orville Wright*, ed. Fred C. Kelly (1951; repr., New York: Da Capo, 1996), 375; "Receipts, Exhibition Department," A. Roy Knabenshue Papers, Archives Division, National Air and Space Museum, Smithsonian Institution (hereafter Knabenshue Papers); "Navy Officer Trains at Wright School," *Aeronautics* 8, no. 5 (May 1911): 159; Orville Wright to Wilbur Wright, 28 May 1911, Family Correspondence: Wright, Orville, 1911, box 6, WBP.

2. Philip Scranton, "Diversity in Diversity: Flexible Production and American Industrialization, 1880–1930," *Business History Review* 65, no. 1 (Spring 1991): 41; Frank H. Russell to Andrew Freedman, 28 July 1911, General Correspondence: Freedman, Andrew, 1911, box 30, WBP; [Wright brothers?] to Andrew Freedman, 5 July 1911, WBP; Wright Company, *Wright Flyers* (New York: Premier Press, 1912), box 108, WBP. "Sales," Wright Company Ledger, 120–21, WCP. There are no extant floor plans for the Wright Company's buildings.

3. Katharine Wright to Wilbur Wright, 23 April, 4 May 1911, box 4, WBP, emphasis in original; Woodbury Pulsifer, ed., *Navy Yearbook* (Washington, DC: Government Printing Office, 1911), 779. In 1911 a navy lieutenant with fewer than five years of service earned $2,400 for shore duty and $2,640 while at sea; after five years his pay rose to $2,640 on shore and $2,904 at sea.

4. David A. Hounshell, *From the American System to Mass Production, 1800–1932: The Development of Manufacturing Technology in the United States* (Baltimore: Johns Hopkins University Press, 1984), 224. For a general study of the business history of Dayton during the era of the Wright Company's existence, see Judith Sealander, *Grand Plans: Business Progressivism and Social Change in Ohio's Miami Valley, 1890–1929* (Lexington: University Press of Kentucky, 1988). Extant Wright Company papers in the Library of Congress and Museum of Flight collections do not suggest that recessions or economic crises directly affected company operations.

5. Barnes to Max Lillie, 27 June 1912, box 2, part 4, WCP; entry for 8 July 1911, Russell diary; Russell to Andrew Freedman, 28 July 1911, General Correspondence: Freedman, Andrew, 1911, box 30, WBP.

6. "First Aeroplane Salesman Here; Booms Wrights," *Kansas City Post*, 27 October 1910, 9; Katharine Wright to Wilbur Wright, 23 April 1911, box 4, WBP; Orville Wright to Wilbur Wright, 23 April 1911, box 4, WBP; Barnes to Frank H. Russell, 25 March, 1 February 1911, WCP; Grover Loening to Alpheus Barnes, 5 January 1914, box 1, 1900–1914, Grover C. Loening Papers, Library of Congress (hereafter Loening Papers); J. I. Clarke to Loening, 6 March 1914, Loening Papers. For examples of Wright Company advertising, see "The Wright Flyer," *Fly*, July 1911, 3; "The Wright Flyer," *Fly* 4, no. 3 (March 1912): 3; "Wright Flyers: 1912 Models," *Aeronautics* 11, no. 3 (September 1912): 98; "A New Wright Flyer," *Fly*, July 1913, 4.

7. Katharine Wright to Milton Wright, 27 July 1909, box 4, WBP.

8. In 1895, George Selden, a Rochester, New York, inventor, received U.S. patent 549,160 for a "road engine" for use in a four-wheeled car. He then used his patent to obtain royalties from other automobile manufacturers (only building autos himself from 1909 to 1912). In 1903, Henry Ford and several other car makers sued Selden over his claims of patent infringement, and in 1911 they won an appellate verdict stating that they had not infringed on Selden's patent in the construction of engines for their autos. Selden's income from royalties collapsed, although his patent would have expired in 1912 in any case.

9. "Wright Company Will Control the Aeroplane," *Dayton Daily News*, 24 November 1909; "Fliers Criticise Wrights," *New York Times*, 5 January 1910, 4; Alpheus Barnes to Orville Wright, 18 June 1913, WCP; "Many Millions Now behind the Wrights to Protect Patents," *Dayton Herald*, 23 November 1909, 1.

10. "Monopoly Not Planned Says Wright," *Dayton Daily News*, 25 November 1909; Tom D. Crouch, "Blaming Wilbur and Orville: The Wright Patent Suits and the Growth of American Aeronautics," in Peter Galison and Alex Roland, eds., *Atmospheric Flight in the Twentieth Century* (Dordrecht: Kluwer Academic, 2000), 291; Glenn H. Curtiss, quoted in Milton Wright to Wilbur Wright, 14 October 1909, box 5, WBP.

11. Wilbur Wright to Orville Wright, 30 June 1911, Family Correspondence, box 7, Wright, Wilbur, 1911, WBP; Wilbur Wright to Orville Wright, 26 May 1911, WBP. The Wrights also filed patent infringement lawsuits in the French and German courts over

patents issued by those countries; while they won in France, the appellate process continued until their patent expired, in 1917. German courts ruled the Wright patent invalid due to prior disclosure.

12. Tom D. Crouch, *The Bishop's Boys: A Life of Wilbur and Orville Wright* (New York: Norton, 1989), 69–85.

13. Wilbur Wright to F. P. Fish, 4 May 1912, General Correspondence: Fish, Richardson and Neave, 1912–13, box 29, WBP; Herbert A. Johnson, "The Wright Patent Wars and Early American Aviation," *Journal of Air Law and Commerce* 69, no. 1 (Winter 2004): 44, 46; Wright Company ledger, WCP, 60–62; "The Aeroplane Patents," *Dayton Daily News*, 2 March 1914, 6; Crouch, *Bishop's Boys*, 451; C. R. Roseberry, *Glenn Curtiss: Pioneer of Flight* (1972; repr., Syracuse, NY: Syracuse University Press, 1991), 338. Johnson also states that the U.S. government did not have the financial resources to provide extensive federal support to aeronautics before the 1913 ratification of the Sixteenth Amendment to the Constitution, which allowed Congress to levy an income tax. The lawsuits involved infringements of the Wrights' flying machine patent issued in May 1906, which expired in 1923. Orville Wright and Wilbur Wright, flying machine, U.S. Patent 821,393, filed 23 March 1903, and issued 22 May 1906.

14. "The Pioneers in Aviation," *Dayton Daily News*, 7 August 1911, 2; "Wright Brothers Recognized as Pioneers of Flying," *Dayton Daily News*, 14 January 1914, 3.

15. "Aeronautics: American Aviation Throttled by Wrights," *Boston Evening Transcript*, 28 February 1914, part 2, p. 4; "Court of Public Opinion," *Aeronautics* 12, no. 3 (March 1913): 90–91; "Another Wright Victory," *New York Times*, 15 January 1914, 8; "Curtiss-Wright Aero Settlement Is Near," *New York Sun*, 12 March 1914, 1; "Wright to Fight for Air Rights," *New York Tribune*, 18 December 1914, 4.

16. Thomas W. McFadden, "Building Industries: Collective Action Problems and Institutional Solutions in the Development of the U.S. Aviation Industry, 1903–1938" (PhD diss., University of Arizona, 1999), 18; John S. Olszowka, "From Shop Floor to Flight: Work and Labor in the Aircraft Industry, 1908–1945" (PhD diss., Binghamton University, 2000), 18; Crouch, "Blaming Wilbur and Orville," 291–93.

17. "Demands Royalties on Aeroplanes," *New York Herald*, 27 February 1914; Roseberry, *Glenn Curtiss*, 339.

18. The Wright Company's ledger contains several pages of slightly conflicting data concerning royalty and license fees; none of the data suggest it received payments significantly in excess of $50,000. See "Royalty Expense," "Royalty Payable," "License Fee Flying," "License Fees Mfrs," and "License Fees (Meets)," Wright Company ledger, 128, 138, 142–44, WCP.

## Chapter 5: World Records for Wright Aviators

1. Alpheus Barnes to Andrew Freedman, 10 March 1911, Wright Corporate Correspondence Ledger, WCP.

2. *Cameron County Press*, 14 September 1911, quoted in William F. Trimble, *High Frontier: A History of Aeronautics in Pennsylvania* (Pittsburgh: University of Pittsburgh Press, 1982), 73. Quimby and passenger William Willard themselves died when they were thrown from the unbelted seats of her in-flight Blériot monoplane at the Third Annual Boston Aviation Meet on 1 July 1912.

3. Charles J. Strobel to Orville Wright, 24 May 1909, General Correspondence: Strobel, Charles J., 1909, box 59, WBP; Francis M. Carroll, *The American Presence in Ulster: A Diplomatic History, 1796–1996* (Washington, DC: Catholic University of America Press, 2005), 109–10; "Paul Knabenshue, U.S. Envoy to Iraq," *New York Times,* 2 February 1942, 15; Crouch, "Chauffeur of the Skies: Roy Knabenshue and the Gasbag Era," *Timeline* 28, no. 2 (April–June 2011): 33, 36–39.

4. Wilbur Wright to Roy Knabenshue, 28 August 1909; Knabenshue to Orville Wright, 25 February 1912, both in General Correspondence: Knabenshue, Roy, box 41, WBP; Wilbur Wright to Knabenshue, 1 March 1910; "Recapitulation of Exhibition Dept., Feb. 14, 1911–Sept. 1, 1911," Wilbur Wright to Knabenshue, 25 October 1911, all in Knabenshue Papers, National Air and Space Museum (NASM); Orville Wright to Wilbur Wright, 30 April 1911, Family Correspondence: Wright, Orville, 1911, box 5, WBP; "Receipts, Exhibition Department," Knabenshue Papers, NASM; "Andre Roosevelt, 83, Explorer and Pioneer in Movies, Is Dead," *New York Times,* 31 July 1962, 27; Katharine Wright to Wilbur Wright, 23 April, 8 June 1911, box 4, WBP.

5. Orville Wright to Wilbur Wright, 29, 31 March 1910, box 6, WBP; "Johnstone Killed by Aeroplane Fall," *New York Times,* 18 November 1910, 2.

6. Tom D. Crouch, *The Bishop's Boys: A Life of Wilbur and Orville Wright* (New York: Norton, 1989), 435; Robert Scharff and Walter S. Taylor, *Over Land and Sea: A Biography of Glenn Hammond Curtiss* (New York: D. McKay, 1968), 205; Ann Deines, "Roy Knabenshue: From Dirigibles to NPS," *CRM* 23, no. 2 (2000): 21.

7. C. R. Roseberry, *Glenn Curtiss: Pioneer of Flight* (1972; repr., Syracuse, NY: Syracuse University Press, 1991), 286, 302; Crouch, *Bishop's Boys,* 429; minutes of the Wright Company executive committee, 11 April 1910, 9 January 1911, WCP; "Beckwith Havens, Early Aviator, 78," [*New York Times*], 9 May 1964, 34. Knabenshue's salary was higher than those of company managers Frank Russell ($3,000 in 1910, $4,000 in 1911) and Grover Loening ($1,800). For salary information, see entry for 29 December 1909, Russell diary; Orville Wright to Loening, 10 July 1913, WCP.

8. Crouch, *Bishop's Boys,* 435; "Brookins in Airship Soars 4,503 Feet," *New York Times,* 18 June 1910, 1; "Goes Up 6,175 Feet in Wright Biplane," *New York Times,* 10 July 1910, 1; "Johnstone Breaks Skyscraping Record," *New York Times,* 1 November 1910, 1; Katharine Wright to Milton Wright, postcard, 26 October 1910 [postmark], box 4, WBP; "Wright Flyers," 6. Flying a Blériot monoplane, J. Armstrong Drexel quickly eclipsed Johnstone's record, reaching 9,970 feet near Philadelphia in November 1910, a week after Johnstone died in a crash near Denver. See "Drexel Wins Record Rising 9,970 Feet," *New York Times,* 24 November 1910, 1.

9. Wilbur Wright to Charles Chandler, 25 January 1912, General Correspondence: Chandler, Charles DeForest, 1912, box 18, WBP; "New American Endurance Record," *Aeronautics* 9, no. 5 (November 1911): 175.

10. Eileen F. Lebow, *Cal Rodgers and the* Vin Fiz: *The First Transcontinental Flight* (Washington, DC: Smithsonian Institution Press, 1989), 71.

11. French Strother, "Flying across the Continent: Rodgers' Trip from Kansas City to Pasadena," *World's Work* 23, no. 4 (February 1912): 405; Wilbur Wright to Orville Wright, 25 September 1911, Family Papers: Correspondence—Wright, Wilbur, July–December

1911, box 6, WBP; Alpheus F. Barnes to C. P. Rodgers, 22 September 1911, WCP. Rodgers's repairs cost the equivalent of nearly $414,000 in 2010 dollars. See Lebow for a detailed study of Rodgers's trip. Rodgers died in a crash of a Wright B five months after landing the *Vin Fiz* in California.

12. "Exhibition Expense at New York Office," Knabenshue Papers, NASM; executive committee minutes, 6 December 1910, WCP; Barnes to M. C. Hoxsey, 18 February 1911, WCP; Crouch, *Bishop's Boys*, 435; "Sixteen Entrants to Race around New York Next Monday," *New York Times*, 11 October 1913, 1.

13. J. A. Adams, "The Man Higher Up," *Advance* 60, no. 2344 (6 October 1910), in Crouch, *Bishop's Boys*, 429; Charles K. Field, "On the Wings of Today," *Sunset* 24, no. 3 (March 1910): 249, in Crouch, *Bishop's Boys*, 429; Barnes to John D. Lindsay, 8 August 1912, WCP.

## Chapter 6: To Change or Not to Change

1. "Orville Wright Demands Royalties on Aeroplanes," *New York Herald*, 27 February 1914; Grover Loening to Baron d'Orcy, 27 March 1914, box 1, 1900–1914, Loening Papers.

2. Kathryn A. Keefer, "The Wright Cycle Shop Historical Report: Greenfield Village," EI 186, Wright Cycle Shop—Simmons Intern Project 2005, Benson Ford Research Center, Dearborn, Michigan, 56.

3. Flaps reduce an airplane's speed by increasing drag and enable pilots to land in shorter distances from a greater angle of descent. A split flap as developed by Wright and Jacobs produces extensive drag and almost no lift. It is uncommon on modern airplanes but was used on the Douglas DC-3.

4. Orville Wright and Wilbur Wright, Flying-Machine, U.S. Patent 1,075,533, filed 10 February 1908, and issued 14 October 1913; Wright and Wright, Flying-Machine, U.S. Patent 987,662, filed 17 February 1908, and issued 21 March 1911; Wright and Wright, Flying-Machine, U.S. Patent 1,122,348, filed 17 February 1908, and issued 29 December 1914; Orville Wright and James M. H. Jacobs, Airplane, U.S. Patent 1,504,663, filed 31 May 1921, and issued 12 August 1924.

5. Glenn H. Curtiss, Aeroplane, U.S. Patent 124,605, filed 18 October 1915, and issued 6 November 1917. For a chart of the number of patent applications filed and patents awarded, see U.S. Patent Office, *Annual Report of the Commissioner of Patents for the Year 1915* (Washington, DC: Government Printing Office, 1916), v. A full list of aviation patents awarded to Curtiss and the Curtiss Motor Company is in the finding aid for the Curtiss-Wright Corporation Archives—Patent Files, National Air and Space Archives, Smithsonian National Air and Space Museum, Chantilly, Virginia, online at http://airandspace.si.edu/research/arch/findaids/pdf/Curtiss-Wright_Patent_Files_Finding_Aid.pdf.

6. "New Wright Model," *Aeronautics* 9, no. 2 (August 1911): 56; "Rebels Take San Ignacio," *New York Times*, 7 February 1911, 1; "The Game from Above," *New York Times*, 5 November 1911, C5.

7. J. Herbert Duckworth, "Aviation Happenings at Home," *Town and Country* 66, no. 1 (27 May 1911): 55; "Society outside the Capital," *Washington Post*, 4 May 1911, 7; Grover C. Bergdoll to "Dear Sirs," n.d. [March 1912], Bergdoll, Grover C., 1912–1913, box 13, WBP;

John Carver Edwards, *Orville's Aviators: Outstanding Alumni of the Wright Flying School, 1910–1916* (Jefferson, NC: McFarland, 2009), 75, 77; Bergdoll to Orville Wright, n.d. [April 1913]; Wright to Bergdoll, 13 April 1913, both in box 13, WBP. The Franklin Institute in Philadelphia now owns Bergdoll's airplane; see http://www.fi.edu/wright/1911/index.html.

8. Edwards, *Orville's Aviators*, 90–94.

9. Herbert A. Johnson, *Wingless Eagle: U.S. Army Aviation through World War I* (Chapel Hill: University of North Carolina Press, 2001), 111. The Wright Company occasionally sold used airplanes; it offered a used Wright B for $3,500 in 1912. See Orville Wright to Wm. Kabitzki [William Kabitzke], 19 June 1912, General Correspondence: Kabitzke, William, 1912, box 39, WBP.

10. Wright Company form letter sent to inquirers, ca. 1911, box 100, Miscellany—Printed Matter, WBP; "The Wright Company School of Aviation," ca. 1914, box 97, Miscellany—Biographical Information, WBP; "The Wright Flying School" (New York: Wright Flying Field, Inc., 1916), 15 (in box 2 of Wright Brothers Collection, Wright Patent Suit folder, Benson Ford Research Center, Dearborn, Michigan); Katharine Wright to Wilbur Wright, 12 May 1911, box 4, WBP. The Wright Company's ledger provides few data about the financing of the Huffman Prairie school or about any profit or loss from the Augusta or Long Island schools.

11. "Millionaire Society Men Learning to Fly in Wright Aviation School at Augusta;" "Wright Company Leases Famous Hempstead Plains Aviation Field," *Flying* 5, no. 4 (May 1916): 152; Edwards, *Orville's Aviators*, 6, 9; "Nearly 200 Winter Students Are Booked," *Aero and Hydro* 5, no. 12 (21 December 1912): 209; Wright Company to Grover C. Bergdoll, 30 January 1912, General Correspondence: Bergdoll, Grover C., 1912–13, box 13, WBP; Wright Company to Howard M. Rinehart, 30 April 1914, box 1, 1900–1914, Loening Papers.

12. "The Aviator—The Superman of Now," Wright Flying Field, Inc., advertisement in *Flying* 5, no. 7 (August 1916): 273.

13. Blanche Scott, quoted in Deborah G. Douglas and Amy E. Foster, *American Women and Flight since 1940* (Lexington: University Press of Kentucky, 2004), 6; John S. Olszowka, "From Shop Floor to Flight: Work and Labor in the Aircraft Industry, 1908–1945" (PhD diss., Binghamton University, 2000), 140.

14. Sales number in Orville Wright to Pliny W. Williamson, telegram, 21 June 1915, General Correspondence: Williamson, Pliny W., 1915, box 66, WBP; E. W. Robischon, "The Evolution of the Wright Airplane," *Aerosphere* 4 (1943): 129; Roger E. Bilstein, *The Enterprise of Flight: The American Aviation and Aerospace Industry* (Washington, DC: Smithsonian Institution Press, 2001), 225; Johnson, *Wingless Eagle*, 111.

15. The Model A, a 1906–9 design, was built by the Wright brothers themselves and not by the Wright Company.

16. Conover, interview by Susan Bennet, 4 February 1967, 53; Orville Wright to G. E. Burroughs, 15 October 1914, box 38, Jones, Rufus B., 1914–17, 1929, WBP; Richard P. Hallion, "The Wright Kites, Gliders, and Airplanes: A Reference Guide," 19 August 2003, www.af.mil/shared/media/document/AFD-051013-002.pdf, 26; "Aero Bibliography: Commercializing the Wright Flyer," *Aero* 1, no. 16 (21 January 1911): 59; Wright Company, *Wright Flyers* (New York: Premier Press, 1912), 4, in box 108, WBP.

No technical drawings or design schematics used by company employees in building Wright Company airplanes are extant; Orville Wright noted in a telegram to aviator George Beatty, who wanted information on how to attach struts to a Wright Company fuselage he purchased from the British government, that he could not help him as relevant "drawings [were] destroyed." See Wright to Beatty, 19 June 1917, Beatty, George W., 1913–17, box 12, WBP.

17. Roger E. Bilstein, "Putting Aircraft to Work: The First Air Freight," *Ohio History* 76, no. 4 (Autumn 1967): 253; "Sales," Wright Company ledger, 121, WCP; "Motorist Buys Wright Plane," *Aero* 1, no. 24 (18 March 1911): 202; A. Timothy Warnock, "From Infant Technology to Obsolescence: The Wright Brothers' Airplane in the U.S. Army Signal Corps, 1905–1915," *Air Power History* 49, no. 4 (Winter 2002): 54.

18. Peter M. Bowers, *Curtiss Aircraft, 1907–1947* (Annapolis: Naval Institute Press, 1979), 40, 41.

19. Ibid., 49; Bartlett Gould, "Burgess of Marblehead," *Essex Institute Historical Collections* 106, no. 1 (January 1970): 21–23.

20. Orville Wright to J. Clifford Turpin, 12 March 1912, General Correspondence: Turpin, J. Clifford, 1912, box 62, WBP; Crouch, *Bishop's Boys*, 457–58; "Price List of Wright Aeroplanes, December, 1913," Wright Aeronautical Company, box 17, Loening Papers.

21. Crouch, *Bishop's Boys*, 458; Wilbur Wright, Orville Wright, and Octave Chanute, *The Papers of Wilbur and Orville Wright, Including the Chanute-Wright Letters and Other Papers of Octave Chanute*, ed. Marvin W. McFarland (1953; repr., Salem, NH: Ayer, 1990), 2:1201, 1203; Grover C. Loening, *Our Wings Grow Faster* (Garden City, NY: Doubleday, Doran, 1935), 50.

22. Crouch, *Bishop's Boys*, 157–59, 447, 449; Milton Wright, *Diaries, 1857–1917* (Dayton: Wright State University, 1999), 748–49; "Final Tribute to World Renowned Inventive Genius," *Dayton Journal*, 2 June 1912, 1; "Entire World Mourns the Loss of Wilbur Wright," *Fly Magazine* 4, no. 9 (July 1912): 9–10; Chip Boyer, "The Door: A Narrative and Commentary Concerning the Funerals of the Wright Brothers and of Other Wright Family Members, 1912–1948" (Dayton: n.p., 2002), 4. Robert Fulton (1765–1815) developed the first commercially successful steamboat; George Stephenson (1781–1848) was known as the Father of Railways and the 4-foot 8½-inch gauge he developed is still the international standard; Alexander Graham Bell (1847–1922), though connected with many fields (including Glenn Curtiss and early aviation), is best remembered for developing the telephone.

23. Crouch, *Bishop's Boys*, 449–50. The First Presbyterian Church no longer exists. The Central Reformed Church purchased the building in 1927. In 1959 Rike's Department Store bought the church and razed it, building a parking garage in its place. A similar situation occurred when Orville Wright died in 1948; his funeral took place downtown at the First Baptist Church, another large structure whose pastor, the Rev. Charles L. Seasholes, was one of the only ministers in Dayton whom Orville respected.

24. Orville Wright to Russell Alger, telegram, 28 May 1912, WBP; Orville Wright to Roy Knabenshue, 9 January 1912; Orville Wright to Turpin, 10 April 1912, both in General Correspondence: Knabenshue, A. Roy, 1912–14, box 41, WBP; Orville Wright to A. A. Merrill, 7 March, 20 April 1912, General Correspondence: Merrill, Albert A., 1912, box 46,

WBP; Orville Wright to Charles D. Chandler, 15 January 1912, General Correspondence; Chandler, Charles DeForest, box 18, WBP; Milton Wright, *Diaries*, 750–56; Johnson, *Wingless Eagle*, 103. Wright Company papers held by the Museum of Flight give the impression that Alpheus Barnes managed the Wright Company's day-to-day operations from New York during the summer of 1912, but they do not provide a complete window into company operations.

25. Milton Wright to Mary A. Wyatt, 18 June 1912, box 5, WBP; Milton Wright, *Diaries*, 750–751; resolution of the Wright Company board of directors, 31 May 1912, WCP; Robert Thum, "The First Jewish Aviator," *Dayton Jewish Observer* 7, no. 6 ( January 2003): 18–19; Katharine Wright to Milton Wright, 15 June 1912, box 4, WBP.

26. Juliette Hennessy, *The United States Army Air Arm, April 1861 to April 1917* (1958; repr., Washington, DC: Office of Air Force History, U.S. Air Force, 1985), 232; Orville Wright to Israel Ludlow, 24 December 1912, box 43, Ludlow, Israel, 1906–18, WBP; "Our Standard Types," Wright Company advertisement, *Aeronautics* 8, no. 2 (August 1913): 79; Henry H. Arnold to Orville Wright, 15 March 1913; Wright to Arnold, 23 March 1913, both in box 11, Arnold, Henry Harley, 1911–13, 1919–31, 1938–47, WBP.

27. Judith Sealander, *Grand Plans: Business Progressivism and Social Change in Ohio's Miami Valley, 1890–1929* (Lexington: University Press of Kentucky, 1988), 45; Peter S. Cajka, "The National Cash Register Company and the Neighborhoods: New Perspectives on Relief in the Dayton Flood of 1913," *Ohio History* 118 (2011): 58, 61, 64.

28. Sealander, *Grand Plans*, 52; "Dayton's Loss, $25,000,000," *New York Times*, 28 March 1913, 1; Orville Wright to Russell A. Alger, 19 April 1913, General Correspondence: Alger, Russell A., 1913, box 10, WBP; "The Wright Co. Contributes Aid," *Dayton Daily News*, 11 April 1913, 1. According to his Winters National Bank check stub books, Wright contributed $125 to the Flood Prevention Fund on 26 June 1913; $375 on 4 April 1914; $375 on 8 April 1916; and $375 on 15 November 1916.

29. "Dayton's Legacy a Gigantic Boom," *New York Times*, 4 May 1913, 12; Wright, Wright, and Chanute, *Papers*, 2:1203; Grover Loening, "The New Wright Model 'E' Single Propeller Biplane and the New Wright Six-Cylinder Motor," *Aircraft* 4, no. 9 (November 1913): 210; Edwards, *Orville's Aviators*, 169; "Elton Aviation Company Organized," *Aeronautics* 15, no. 7 (15 October 1914): 104. Elmer Sperry's gyroscopic stabilizer, introduced in 1914, became the stabilizer from which future stabilizers derived; Wright's was never used commercially.

30. Wright, Wright, and Chanute, *Papers*, 2:1205; Warnock, "Infant Technology," 54; Hennessy, *Army Air Arm*, 117.

31. "The Model 'CH' Wright Waterplane," *Flight* 5, no. 36 (6 September 1913): 978; "New Model 'CH' Wright," *Aeronautics* 13, no. 1 ( July 1913): 11; Dwight D. Eisenhower to Chief, Motor Transport Corps, "Report on Trans-Continental Trip," 3 November 1919, www.fhwa.dot.gov/infrastructure/convoy.cfm; Grover Loening, "The Wright Company's New Hydro-Aeroplane Model C-H," *Aircraft* 4, no. 7 (September 1913): 152; Wright Company advertisement, *Aeronautics* 13, no. 2 (August 1913): 79; Hallion, "Wright Kites," 42; "Price List of Wright Airplanes, December, 1913," box 108, Miscellany, 1910–1916, n.d., WBP; Gordon Swanborough and Peter M. Bowers, *United States Navy Aircraft since 1911* (Annapolis: Naval Institute Press, 1990), 536.

32. "Men Will Soon Fly with Absolute Safety, Says Wright," *New York Tribune*, 16 November 1913, 3; "Wright Model G Aeroboat," Wright Company advertisement, *Aeronautics* 13, no. 5 (November 1913): 185; Hallion, "Wright Kites," 49; "New Aeroboat to Go 67 Miles in an Hour," *New York Sun*, 11 August 1913, 2; Grover C. Loening, *Takeoff into Greatness: How American Aviation Grew So Big So Fast* (New York: Putnam, 1968), 57; Herman Schier to Grover Loening, 13 December 1914, box 1, 1900–1914, Loening Papers; "The New Wright Control," *Flight* 6, no. 10 (7 March 1914): 240. For descriptions of the two aircraft, see "The 100 H.P. Curtiss Flying Boat," *Flying* 6, no. 15 (11 April 1914): 384–87; "The 60 H.P. Wright Aero-Boat," *Flying* 6, no. 2 (17 January 1914): 56–58.

33. The crash of the ZR-2/R38, a British-built airship purchased by the U.S. Navy, killed forty-four of the forty-nine men aboard the airship, including almost all of the navy's rigid airship personnel.

34. "Wright Licenses Granted," *Aeronautics* 14, no. 10 (30 May 1914): 147; Grover C. Loening, "The New Wright Aeroboat Type 'G,'" *Aeronautics* 14, no. 11 (15 June 1914): 171; "Airmen Wait for Call to Serve in Mexico," *New York Sun*, 23 April 1914, 6; "Naval Orders," *Washington Post*, 20 August 1914, 7 (which misspells Maxfield's name as Maxifold); "Classified Advertising," *Aircraft* 5, no. 5 (July 1914): 343; Tom Crouch, e-mail message to author, 14 February 2012; "Sales," Wright Company ledger, WCP, 146. In 2002 the National Air and Space Museum loaned Hall's Wright G to the National Park Service for display at the Dayton Aviation Heritage National Historical Park.

35. Hallion, "Wright Kites," 53; "The New Wright Biplane," *Flight* 6, no. 47 (20 November 1914): 1133; Wright, Wright, and Chanute, *Papers*, 2:1207–8; John H. Morrow, *The Great War in the Air: Military Aviation from 1909 to 1921* (Washington, DC: Smithsonian Institution Press, 1993), 75–81, 113–23; "Bonney Goes to Fly in Mexico," *Aerial Age Weekly* 1, no. 6 (26 April 1915): 128.

36. Milton Wright, *Diaries*, 800; "Navy Opens Bids for Nine Hydros," *Aeronautics* 16, no. 1 (15 March 1915): 4; Hallion, "Wright Kites," 55.

37. Hallion, "Wright Kites," 57; Wright, Wright, and Chanute, *Papers*, 2:1208–10; "The New Wright Tractor Biplane—Type L," *Flight* 8, no. 32 (10 August 1916): 664.

### Chapter 7: Turning Buyer Attention the Company Way

1. See "The Wright Flyer," Wright Company advertisement, *Town and Country* 66, no. 11 (27 May 1911): 48; "The Wright Flyer," Wright Company advertisement, *Town and Country* 66, no. 22 (12 August 1911): 22.

2. Lyell D. Henry, Jr., *Zig-Zag-and-Swirl: Alfred W. Lawson's Quest for Greatness* (Iowa City: University of Iowa Press, 1991), 61; "Wright Flyers," Wright Company advertisement, *Aeronautics* 7, no. 1 (July 1910): 28; "The Wright Flyer," Wright Company advertisement, *Aeronautics* 7, no. 5 (November 1910): 156; "The Wright Flyer," Wright Company advertisement, *Aeronautics* 8, no. 5 (May 1911): 176.

3. "The Wright Flyer," Wright Company advertisement; "Lumina Aeroplane Cloth," B. F. Goodrich advertisement, both in *Aeronautics* 10, no. 4 (April 1912): 136.

4. Alfred Lawson to Grover Loening, 24 July 1913; Ernest Jones to Loening, 21 August 1913, both in box 1, 1900–14, Loening Papers. Grover Loening's articles included "The New Wright Aeroboat Type 'G,'" *Aeronautics* 14, no. 11 (15 June 1914): 171; "The Wright

Company's New Hydro-Aeroplane Model C-H," *Aircraft* 4, no. 7 (September 1913): 152; "The New Wright Model 'E' Single Propeller Biplane and the New Wright Six-Cylinder Motor," *Aircraft* 4, no. 9 (November 1913): 210.

5. "The Wright Company," Wright Company advertisement, *Aeronautics* 14, no. 4 (28 February 1914): 64; Alpheus Barnes to Loening, 3 December 1913, box 1, 1900–14, Loening Papers; "The New Wright Aeroplanes," Wright Company advertisement, *Aeronautics* 14, no. 11 (15 June 1914): 173; "Wright Aeroplanes," Wright Company advertisement, *Aerial Age Weekly* 1, no. 26 (13 September 1915): 611; "Wright News," *Aerial Age Weekly* 1, no. 14 (21 June 1915): 319.

6. For Dayton, see "Wright Aeroplanes," Wright Company advertisement, *Aerial Age Weekly* 1, no. 26 (13 September 1915): 611; for Augusta, see "The Wright Flying School," Wright Company advertisement, *Aerial Age Weekly* 2, no. 16 (3 January 1916): 370; for Hempstead Plains, see "The Wright Flying School," Wright Company advertisement, *Aerial Age Weekly* 3, no. 16 (3 July 1916): 486.

7. "Mr. Reader! Why Are You Interested in Aviation?" Curtiss Exhibition Company advertisement, *Aircraft* 2, no. 10 (December 1911): 329, and *Popular Mechanics* 17, no. 1 (January 1912): 147; Mary Seelhorst, "*Popular Mechanics*: 90 Years," *Popular Mechanics* 169, no. 2 (February 1992): 83; "Curtiss Features," Curtiss Aeroplane Company advertisement, *Aeronautics* 8, no. 5 (May 1911): 179, and *Aircraft* 2, no. 3 (May 1911): 90.

8. "You Can Now Fly with Safety!" Curtiss Aeroplane Company advertisement, *Aeronautics* 11, no. 4 (October 1912): 121; "Results Tell the Story," Curtiss Aeroplane Company advertisement, *Aeronautics* 13, no. 6 (December 1913): cover; "Some Recent Curtiss Aeroplanes," Curtiss Aeroplane Company advertisement, *Aeronautics* 15, no. 6 (30 September 1914): cover; "Curtiss Facilities," Curtiss Aeroplane Company advertisement, *Aeronautics* 16, no. 7 (15 June 1915): 109; "Curtiss Military Tractor," Curtiss Aeroplane Company advertisement, *Aerial Age Weekly* 2, no. 1 (20 September 1915): 2; "Reliable—Efficient—Durable," Curtiss Aeroplane Company advertisement, *Aerial Age Weekly* 2, no. 9 (15 November 1915): 194.

9. "Burgess Biplanes," Burgess Company and Curtis advertisement, *Aeronautics* 8, no. 3 (March 1911): 88, and *Town and Country* 66, no. 11 (27 May 1911): 68; "Burgess Aeroplanes," Burgess Company and Curtis advertisement, *Aircraft* 2, no. 8 (October 1911): 287; "New American Record," Burgess Company and Curtis advertisement, *Aeronautics* 12, no. 4 (April 1913): 123; "One of the Burgess Flying Boats Built for the U.S. Navy," Burgess Company and Curtis advertisement, *Aircraft* 4, no. 10 (December 1913): 217; "Our Aeroplanes and Hydroplanes Have Become the American Standard," Burgess Company and Curtis advertisement, *Aeronautics* 12, no. 2 (February 1913): 43; "One of the Burgess Flying Boats Built for the U.S. Navy," Burgess Company and Curtis advertisement, *Aircraft* 4, no. 12 (February 1914): 273; "Burgess-Dunne Military Aeroplane," Burgess Company advertisement, *Aeronautics* 15, no. 8 (30 October 1914 [published 11 February 1915]): 121; Bartlett Gould, "Burgess of Marblehead," *Essex Institute Historical Collections* 106, no. 1 (January 1970): 25.

10. "Sales," Wright Company ledger, 40–41, 120–21, 145–47; "Dividend Acct.," 129, both in WCP; Winters National Bank check stub books, 20 January 1911–8 June 1912, 26 October 1912–16 December 1913, series 3, subseries 2, box 13, WBP.

11. John S. Olszowka, "From Shop Floor to Flight: Work and Labor in the Aircraft Industry, 1908–1945" (PhD diss., Binghamton University, 2000), 23–25; "Curtiss to Move His Plant," *New York Times*, 6 December 1914, 1. Olszowka suggests that the various Curtiss companies were legally separate to protect them from any patent-related court-ordered seizure by the Wright Company.

12. Gould, "Burgess of Marblehead," 22–29; "Aeroplane Factory Moves," *Los Angeles Times*, 4 January 1913, I15; "Glenn Martin Gets Offer for Output," *Santa Ana Blade*, 3 May 1915.

### Chapter 8: Managing the Wrights' Company

1. Wright Company, *Wright Flyers* (New York: Premier Press, 1912), 3; Wright brothers to A. A. Merill [*sic*], 7 December 1909, General Correspondence: Merrill, Albert A., box 46, WBP.

2. Entries for 9 December 1909; 12 July, 8 August 1910; 7, 8 January 1911, Russell diary; Robert Friedel, *Zipper: An Exploration in Novelty* (New York: Norton, 1994), 75; Wright brothers to Russell, 4 January 1910, box 55, WBP.

3. Tom Russell, interview by Susan Bennet, 24 April 1967, 33; Russell A. Alger to Wilbur Wright, 4 October 1911, General Correspondence: Alger, Russell, A., 1910–11, box 10, WBP; Milton Wright, *Diaries*, 737.

4. Entries for 12 July, 8 August 1910, Russell diary; Wilbur Wright to Knabenshue, 1 March 1910, Knabenshue Papers; Knabenshue to Wilbur Wright, 15 February 1911, General Correspondence: Knabenshue, Roy, box 41, WBP; Orville Wright to Knabenshue, 27 March 1911, Knabenshue Papers; Katharine Wright to Wilbur Wright, 23 April 1911, box 4, WBP.

5. Wilbur Wright to Frank H. Russell, 4 January 1910, General Correspondence: Russell, Frank H., 1909–12, 1927–38, 1947, box 55, WBP; Orville Wright to Wilbur Wright, 28 May 1911, Family Correspondence, box 6, Wright, Orville, 1911, WBP.

6. Entry for 13 July 1911, Russell diary; Frank H. Russell to Orville Wright, 14 July 1911, General Correspondence: Russell, Frank H., 1909–12, 1927–38, 1947, box 55, WBP; Wilbur Wright to Orville Wright, 26 May 1911, Family Correspondence, box 7, Wright, Wilbur, March–June 1911, WBP.

7. Frank H. Russell to Freedman, 28 July 1911, General Correspondence: Freedman, Andrew A., 1911, box 30, WBP; entries for 17, 19, 20, 26 July 1911, Russell diary.

8. Freedman to Orville Wright, 28 July 1911; Freedman to Wright, telegram, 31 July 1911, both in General Correspondence: Freedman, Andrew A., 1911, box 30, WBP; Milton Wright, *Diaries*, 698.

9. John S. Olszowka, "From Shop Floor to Flight: Work and Labor in the Aircraft Industry, 1908–1945" (PhD diss., Binghamton University, 2000), 172; Wilbur Wright to Orville Wright, 25 September 1911, box 7, Wright, Wilbur, July–December 1911, WBP.

10. Frank H. Russell to Orville Wright, telegram, 26 July 1911; Orville Wright to Russell, telegram, 29 July 1911, both in box 55, General Correspondence: Russell, Frank H., 1909–12, 1927–38, 1947, WBP; "Sales," Wright Company ledger, 121, WBP.

11. Alpheus Barnes to F. H. Russell, 5, 7, 25 August, 7 September 1911, WCP; entries for 6, 19, 20, 27 September, 12 October 1911, Russell diary; Russell to Wilbur Wright, 31

October 1911, box 55, General Correspondence: Russell, Frank H., 1909–12, 1927–38, 1947, WBP; Katharine Wright to Milton Wright, telegram, 19 October 1911, box 4, WBP; Russell Alger to Wilbur Wright, 4 October 1911, General Correspondence: Alger, Russell A., 1910–11, box 10, WBP.

12. "F. H. Russell Dead; Leader in Aviation," *New York Times*, 5 August 1947, 23. See Frank H. Russell, "Burgess Flying Boat Solves Warping Difficulties," *Aircraft* 4, no. 4 (June 1913): 79; "The Burgess Flying Boat Built for Robert J. Collier," *Aircraft* 4, no. 7 (September 1913): 164; "Description of the Burgess-Dunne Hydro-aeroplane," *Aircraft* 5, no. 2 (April 1914): 296–98.

13. Wilbur Wright to Orville Wright, 28 September 1911, Family Correspondence: Wright, Wilbur, July–December 1911, box 7, WBP; Winters National Bank check stub books, multiple entries for Feight between 18 May 1912 and 29 June 1912, Wright Brothers Papers, Wright State University.

14. Loening to Orville Wright, 2 June, 5 July 1912; Orville Wright to Loening, 8 July 1912; Willis McCornick to Loening, 9 November 1911, Loening Papers; Loening, *Our Wings Grow Faster*, 12. Loening remained enrolled at Columbia into 1911, when he earned a civil engineering degree.

15. Loening, *Our Wings Grow Faster*, 21–30, 44; "Martin Suit Dismissed," *New York Times*, 11 June 1940, 18; Loening to Orville Wright, 22 June 1913; Orville Wright to Loening, 7 July 1913, both in box 43, General Correspondence: Loening, Grover C., 1910–13, WBP.

16. Orville Wright to Grover Loening, 10 July 1913, WCP; H. H. Arnold to Loening, 29 July 1913; Roy C. Kirtland to Loening, 2 August 1913; Glenn H. Curtiss to Loening, 4 August 1913; Alfred W. Lawson to Loening, 21 July 1913; Loening to Henry Woodhouse, 3 October 1913; Ernest L. Jones to Loening, 21 August 1913, all in box 1, 1900–1914, Loening Papers; Wright Company executive committee minutes, 28 January 1914, WCP.

17. Loening, *Our Wings Grow Faster*, 43, 42; Loening, *Takeoff into Greatness: How American Aviation Grew So Big So Fast* (New York: Putnam, 1968), 55; Loening to Lt. T. DeWitt Milling, 8, 23 September 1913, box 1, 1900–1913, Loening Papers.

18. Loening to W. I. Chambers, 19 August 1913; Chambers to Loening, 17 October 1913; Oscar Brindley to Loening, 15 December 1913; Brindley to Orville Wright, 15 December 1913, all in box 1, 1900–1913, Loening Papers.

19. Tom Russell, interview, 33; Loening to American Service Company, 6 August 1913; Loening to Herman Schier, 19 August 1913, both in box 1, 1900–1914, Loening Papers. Tom Russell enjoyed working for Frank Russell.

20. Loening to Henry Woodhouse, 8 September 1913; Woodhouse to Loening, 29 September 1913; Loening to Woodhouse, 3 October 1913, all in box 1, 1900–1914, Loening Papers; Loening, quoted in Johnson, *Wingless Eagle*, 104; Loening, *Our Wings Grow Faster*, 44, 45; Loening, *Takeoff into Greatness*, 54–57.

21. Loening, *Takeoff into Greatness*, 54–55, 59, 62, emphasis in original. Loening discussed his resignation with Wright on 28 May and formalized it in a letter on 2 June 1914. See Loening to Wright, 2 June 1914, box 1, 1900–1914, Loening Papers.

22. Warnock, "Infant Technology," 55; Oscar Brindley to Loening, 15 December 1913, box 1, 1900–1914, Loening Papers.

23. C. P. Nellis to Loening, 15 December 1914, box 1, 1900–1914, Loening Papers. Nellis remained in Dayton, where he later worked for the Dayton-Wright Airplane Company.

24. U.S. Census Office, *Tenth Census, 1880, Jersey City, Hudson County, New Jersey*, s.v. "Thadeus Barnes," Heritage Quest, www.heritagequestonline.com; "New Jersey Deaths and Burials, 1720–1988," index, FamilySearch, https://familysearch.org/pal:/MM9.1.1/FZ6F-TWL, Thaddeus C. Barnes, 1847, citing reference Reg. of Deaths, FHL microfilm 589833.

25. Loening, *Our Wings Grow Faster*, 44; Loening, *Takeoff into Greatness*, 54–55, 61–62.

26. Frank Russell notes that Barnes was in Dayton by 14 July; Barnes's date of departure for New York is unclear, though he sent Russell a telegram from Asbury Park, New Jersey, on August 12. Entries for 14 July, 12 August 1910, Russell diary.

27. Entries for 5, 15 January, 15 June 1910; 11 July 1911, Russell diary, emphasis in original; Wilbur to Andrew Freedman, 5 July 1910, Andrew Freedman to Wilbur Wright, 8 July 1910, both in box 30, General Correspondence: Freedman, Andrew A., 1909–1910, WBP; Milton Wright, *Diaries*, 718, 745, 773, 781, 783, 788.

28. Orville Wright to Wilbur Wright, 11 June 1911, box 6, WBP; Wilbur Wright to Orville Wright, 25 September 1911, box 7, WBP; "Airship Outraces Auto," [*Salt Lake Tribune*], 31 May 1911, 11; "Minutes of a Special Meeting of the Board of Directors of the Wright Company," 9 October 1915, WCP.

29. U.S. Department of Commerce, Bureau of the Census, *Fourteenth Census, 1920, Jefferson Township, Morris County, New Jersey*, s.v. "Sarah Barnes," Heritage Quest, www.heritagequestonline.com; registration card, Alpheus Fayette Barnes, United States World War I draft registration cards, 1917–18, East Orange City, New Jersey, www.familysearch.org.

## Chapter 9: "It Is Something I Have Wanted to Do for Many Months"

1. Michael Eckert, "Strategic Internationalism and the Transfer of Technical Knowledge: The United States, Germany, and Aerodynamics after World War I," *Technology and Culture* 46, no. 1 (January 2005): 107.

2. "Theodore P. Shonts," *New York Times*, 22 September 1919, 10; Wright to Pliny Williamson, 14 April 1914, General Correspondence: Williamson, Pliny W., 1914, box 65, WBP; Milton Wright, *Diaries*, 303, 815. Cornelius Vanderbilt, who did not express an opinion on McCombs, was also active in Republican politics as a delegate to the 1900 New York state party convention and as a member of the civil service commission of New York City mayor Seth Low, a Republican, who served from 1902 to 1903.

3. Wright to Williamson, 14 April 1914, box 65, General Correspondence: Williamson, Pliny W., WBP; "Wright Would Quit Airplane Selling," *New York Times*, 31 May 1914, 11; Wright to Frederick M. Alger, 15 April 1914, box 9, General Correspondence: Alger, Frederick M., WBP; Milton Wright, *Diaries*, 787; Wright to Robert J. Collier, 29 April, 24 November 1914, 1 October 1915, box 20, General Correspondence, Collier, Robert J., 1910–18, WBP; Wright to Williamson, 11 November 1914; Williamson to Wright, 24 August 1915; Williamson to Wilbur and Orville Wright, 16 May 1908, box 65, WBP; Loening *Takeoff into Greatness*, 63; Wright to Robert Collier, 24 November 1914. Two hundred thousand dollars in 1914 was the equivalent of approximately $4.5 million in 2010. Williamson (1876–1958) graduated with a BA from Oberlin in 1899, a year after Katharine Wright. In the 1900 U.S. Census, he is enumerated twice, both times as a lawyer: on 6 June, at his parents' residence, 122 South Broadway Street in Dayton, two blocks west of the Wright home, at 7 Hawthorn Street; on 28 June, boarding at 239 Warm Springs Avenue

in Boise, Idaho. See U.S. Census Office, *Twelfth Census, 1900, Dayton, Fifth Ward, Montgomery County, Ohio,* s.v. "Pliny Williamson"; U.S. Census Office, *Twelfth Census, 1900, Boise Precinct 1, Ada County, Idaho,* s.v. "Pliny Williamson," Heritage Quest, www.heritagequestonline .com. See also "P. W. Williamson, Legislator, Dies," *New York Times,* 22 October 1958, 35.

4. Tom D. Crouch, *The Bishop's Boys: A Life of Wilbur and Orville Wright* (New York: Norton, 1989), 485–87; Lorin Wright to Orville Wright, 3, 4 June 1915 and telegram, 3 June 1915, both in Family Correspondence: Wright, Lorin, 1911–15, box 5, WBP.

5. "Minutes of a Special Meeting of the Board of Directors of the Wright Company," 9 October 1915, box 2, WCP; Wright to Williamson, 11 November 1914, box 65, General Correspondence: Williamson, Pliny W., 1914, WBP.

6. "R. J. Collier Dies at Dinner Table," *New York Times,* 9 November 1918, 13; "Andrew Freedman Dies of Apoplexy," *New York Times,* 5 December 1915, 19; "T. P. Shonts Dies at His Home Here," *New York Times,* 21 September 1919, 1; "Russell A. Alger Dies of Pneumonia," *New York Times,* 27 January 1930, 19; "De Lancey Nicoll, Noted Lawyer, Dies," *New York Times,* 1 April 1931, 24. U.S. Department of Commerce, Bureau of the Census, *Fourteenth Census, 1920, Jefferson Township, Morris County, New Jersey,* s.v. "Sarah Barnes"; U.S. Department of Commerce, Bureau of the Census, *Fourteenth Census, 1920, Los Angeles Township, Los Angeles, California,* s.v. "Alpheus F. Barnes."

7. "Man Who Built Wrights' Plane Engine Dies," *Los Angeles Times,* 31 January 1956, 12; U.S. Department of Commerce, Bureau of the Census, *Fifteenth Census, 1930, Millcreek Township, Hamilton County, Ohio,* s.v. "Arthur A. Gaible"; *Precinct P, Hamilton County, Ohio,* s.v. "Rufus B. Jones"; *Tenth Ward, Dayton, Montgomery County, Ohio,* s.v. "Louis C. Luneke"; *Fourth Ward, Dayton, Montgomery County, Ohio,* s.v. "Frank T. Whipp."

8. Russell interview, Conover interview, and Kreusch interview, all held by the Roesch Library at the University of Dayton; "R. W. Elliot, Wright Aide, Dies at 91," *Dayton Daily News,* 9 January 1963, 32; "Frank T. Whipp," *Dayton Daily News,* 6 October 1942, 17; "Louis C. Luneke," *Dayton Daily News,* 4 May 1963, 3; "Thomas W. Russell," *Dayton Daily News,* 15 June 1970, 35; "Ida Holdgreve," *Dayton Daily News,* 29 September 1977, 35.

9. "Aeroplane Industry's Great Growth," *Aerial Age Weekly* 3, no. 16 (3 June 1916): 473; Hennessy, *Army Air Arm,* 134; Crouch, *Bishop's Boys,* 484–85. The lawsuit, *Wright Company v. Curtiss Aeroplane Company,* was filed in equity in the U.S. District Court in the Western District of New York.

10. Orville Wright to Roy Knabenshue, 10 November 1915, General Correspondence: The Wright-Martin Aircraft Corporation, box 67, WBP; "Wright Denies Selling Out to Eastern Company," *Dayton Journal,* 3 May 1914, 1, 10. By 1915, Wright's European assets were minimal. The German Wright Company—Flugmaschine Wright GmbH—went bankrupt in 1914; contracts with Short Brothers in Britain and the Compagnie générale de navigation aérienne in France were not profitable and produced few airplanes.

11. Orville Wright to Pliny Williamson, 23 December 1914, box 65, General Correspondence: Williamson, Pliny W., 1914, WBP; William J. Hammer to Orville Wright, 28 June 1915, box 33, General Correspondence: Hammer, William J., 1915–18, WBP; Williamson to Wright, 20 November 1914, box 65, General Correspondence: Williamson, Pliny W., 1914, WBP; Wright to Williamson, 6 October 1914, WBP; Wright to Williamson, 14 September 1914, WBP; Milton Wright, *Diaries,* 790, 793. For a time in 1914

the company's New York telephone was disconnected for nonpayment of bills; see Williamson to Wright, 11 November 1914, WBP.

12. Wright to Williamson, 11 November 1914; Williamson to Wright, 14, 20 November 1914, both in General Correspondence: Williamson, Pliny W., 1914, box 65, WBP; "Daniels Ready to Sign for Hydroaeroplanes," *Washington Times*, 17 April 1915, 14; "Nation's First Dirigible Ready for Maiden Cruise," *New York Tribune*, 11 August 1915, 1–2; Wright to Williamson, 3 April 1915; Williamson to Wright, 26 May 1915, box 65, WBP; Milton Wright, *Diaries*, 801; Williamson to Wright, 12 June 1915; Williamson to Wright, telegram, 29 June 1915, box 65, WBP.

13. Orville Wright, confirmation of telegram to Mr. F. P. Fish, 16 September 1915, box 29, Fish, Richardson, and Neave, 1915–16, WBP; Williamson to Wright, 14 August 1915, box 66, WBP; Williamson to Wright, 27 September 1915, box 66, WBP; Wright to Collier, 1 October 1915, box 20, WBP; Williamson to Wright, 5 October 1915, box 66, WBP; Williamson to Katharine Wright, 5 October 1915, box 66, WBP; "Aeroplane Rights Sold by O. Wright," *New York Times*, 14 October 1915, 8; "Wright Glad to Be Free," [*New York Times*], 14 October 1915, 8. The Thomas Aeromotor company (later the Thomas-Morse Aircraft Corporation) was an Ithaca, New York–based airplane builder. The Consolidated Aircraft Corporation purchased it in 1929 and closed it in 1934.

14. "Syndicate Takes Over Wright Aeroplane Plant," *Wall Street Journal*, 15 October 1915, 2; "Wright's Patents and Factory Are Sold in the East," *Dayton Herald*, 13 October 1915, 16; "Orville Wright Will Retire from Business," *Philadelphia Inquirer*, 29 August 1915, 3; "Fifteen Aeroplanes a Day for Europe," *Aerial Age Weekly* 2, no. 3 (4 October 1915): 53; "Million and Half Involved in Sale of Wright Plant," *Dayton Journal*, 14 October 1915, 14.

15. The body of Williamson's lawsuit against Wright is a mystery; I have been unable to locate briefs, opinions, or any other case records save a receipt for the settlement money signed on behalf of Williamson by his lawyers. Williamson last communicated with Lorin Wright in November 1915 and accepted $20,000 from Orville Wright in May 1916. What specifically transpired in the intervening six months, or whether Williamson's lawsuit against Wright actually went to trial, is unclear.

16. "Says O. Wright Got $500,000 for Stock," *New York Times*, 16 November 1915, 7; Williamson to Orville Wright, 27 September 1915; Lorin Wright to Williamson, 11 November 1915, both in box 66, WBP; Williamson to Orville and Wilbur Wright, 4 January 1910, box 65, WBP; Williamson to "My dear Boys" (Wilbur and Orville Wright), 17 November 1909, box 65, WBP; receipt, *Pliny W. Williamson v. Orville Wright*, U.S. District Court, Southern District of New York, 6 May 1916, box 66, WBP. The par value of Wright Company stock was $100 per share, giving the stock held by Wright at the time of the sale a book value of $776,000. A sale figure of $1.5 million in 1915 seems especially high in light of the inventory of the frugal Wright's probated estate after his 1948 death, which placed his assets at slightly more than $1 million; see "Inventory Puts Wright Estate at $1,067,105.73," *Dayton Herald*, 18 March 1948.

17. Winters National Bank check stub book, 7 September 1915–8 July 1916, WBP; Milton Wright, *Diaries*, 809; "Sales Acct.," Wright Company ledger, p. 82, WCP; Orville Wright and James M. H. Jacobs, airplane, U.S. Patent 1,504,663, filed 31 May 1921, and granted 12 August 1924; Woodrow Wilson to Orville Wright, 29 January 1920, box 66, Wilson, Woodrow,

1920, WBP; Alex Roland, *Model Research: The National Advisory Committee for Aeronautics, 1915–1958* (Washington, DC: Scientific and Technical Information Branch, National Aeronautics and Space Administration, 1985), 664. A $315 weekly payroll over a fifty-two-week year equals $16,380; if each worker earned the 1915 national nonfarm average of $687 annually, that would suggest a final payroll of approximately twenty-four employees.

18. "Simplex Motor with Wright Aeroplane Co.," *Aerial Age Weekly* 2, no. 11 (29 November 1915): 248; "Wright Company's Winter School in the South," *Aerial Age Weekly* 2, no. 14 (20 December 1915): 327; "Wright Aeroplane Plans a Big Output," *Wall Street Journal*, 30 December 1915, 6; "Aero Corporations Merge," *New York Times*, 8 August 1916, 13; "News and Notes of the Automobile Trade," *New York Times*, 13 August 1916, xx8.

19. "Wright Company, New York, and Glenn L. Martin Company, Los Angeles, Merge," *Aerial Age Weekly* 3, no. 22 (14 August 1916): 653; "Largest Aeroplane Co. in the World Will Be Represented in Dayton," *Dayton Daily News*, 17 August 1916, 1; Orville Wright to Wright Company, transcription of telegram, 19 February 1917, Wright-Martin Aircraft Corp., box 67, WBP.

20. Manufacturers Aircraft Association, *Aircraft Yearbook, 1920* (New York: Doubleday, Page and Company, 1920), 241; Roseberry, *Glenn Curtiss*, 333; Curtiss Wright Corporation, "Facilities," http:curtisswright.com/facilities.asp. Neither Orville Wright nor Glenn Curtiss had any meaningful role in their respective companies at the time of the merger in 1929.

*Epilogue*

1. "M'Donnell Douglas Assails Trust Suit," *New York Times*, 1 April 1972, 31.

2. "Wright-Curtiss Litigation," *Aeronautics* 12, no. 3 (March 1913): 85.

3. "It Certainly Is a Darling," *Motor West* 27, no. 2 (1 May 1917): 26; "Aeroplane Plant Is Acquired by Auto Co.," *Dayton Herald*, 22 March 1917, 1, 4.

4. Harold E. Talbott, Jr., served as chairman of North American Aviation and as a board member of Trans World Airlines in the 1930s and was secretary of the air force from 1953 to 1955, coordinating the selection of Colorado Springs as the site of the Air Force Academy.

5. Tom D. Crouch, *The Bishop's Boys: A Life of Wilbur and Orville Wright* (New York: Norton, 1989), 470; "Airplane Workers Exempt," *New York Times*, 19 January 1918, 4; Frank S. Adams, "Women in Democracy's Arsenal," *New York Times*, 19 October 1941, 29.

6. Fay L. Faroute, ed., *Aircraft Year Book, 1919* (New York: Manufacturers Aircraft Association, 1919), 130; Dayton-Wright Airplane Company, *Some Interesting Facts and Photographs* (Dayton: Dayton-Wright Airplane Company, [ca. 1919]), 9; "End Patent Wars of Aircraft Makers," *New York Times*, 7 August 1917, 5. Dayton-Wright's Plant 1, its principal factory, was in Moraine; Plant 2 was in Miamisburg. Neither still exists.

7. Manufacturers Aircraft Association, *Aircraft Year Book, 1920* (New York: Doubleday, Page and Company, 1920), 197–200; Stuart W. Leslie, *Boss Kettering* (New York: Columbia University Press, 1983), 96; "The Camera Looks at Inland's Progress . . . ," *Inlander*, Special 15th Anniversary Edition (1937): 8, in Harvey Geyer Collection, file 4, Dayton History Archive Center; Omnibus Public Land Management Act of 2009, Pub. L. no. 111–11, 123 Stat. 991 (2009).

# BIBLIOGRAPHY

## Archival Sources

Bowd Osborne Photographs and Color Slides, Marblehead Historical Commission, Marblehead, Massachusetts

Dayton History Archive Center

Dayton Metro Library

Library of Congress, Washington, DC

Museum of Flight, Seattle

Smithsonian Institution, National Air and Space Museum, Archives, Chantilly, Virginia

University of Wyoming, American Heritage Center, Laramie

Wright State University, Special Collections and Archives, Fairborn, Ohio

## Newspapers

*Atlanta Constitution*
*Boston Evening Transcript*
*Chicago Daily Tribune*
*Dayton Daily News*
*Dayton Herald*
*Dayton Journal*
*Dayton Journal Herald*
*Detroit Free Press*
*Evening Post* (Wellington, New Zealand)
*Fort Worth Star-Telegram*
*Hartford Courant*
*Kansas City Post*
*Los Angeles Times*
*New York Herald*
*New York Sun*
*New York Times*
*New York Tribune*
*New York World*
*Philadelphia Inquirer*
*Salt Lake Tribune*
*Santa Ana Evening Blade*
*The Times* (London, England)
*Wall Street Journal*
*Washington Post*
*Washington Times*

## Books and Articles

"Aero Bibliography: Commercializing the Wright Flyer." *Aero* 1, no. 16 (21 January 1911): 59.

"Aeroplane Industry's Great Growth." *Aerial Age Weekly* 3, no. 16 (3 June 1916): 473.

Auerbach, Joseph S. *De Lancey Nicoll: An Appreciation.* New York: Harper and Brothers, 1931.

Becker, Carl. "Mill, Shop and Factory: The Industrial Life of Dayton, Ohio, 1830–1900." PhD dissertation, University of Cincinnati, 1971.

———. "A 'Most Complete' Factory: The Barney Car Works, 1850–1926." *Cincinnati Historical Society Bulletin* 31, no. 1 (Spring 1973): 47–69.

Bednarek, Janet. *Reconsidering a Century of Flight.* Chapel Hill: University of North Carolina Press, 2003.

Biddle, Wayne. *Barons of the Sky: From Early Flight to Strategic Warfare; The Story of the American Aerospace Industry.* New York: Henry Holt, 1991.

Biggs, Lindy. *The Rational Factory: Architecture, Technology, and Work in America's Age of Mass Production.* Baltimore: Johns Hopkins University Press, 1996.

Bilstein, Roger E. *The Enterprise of Flight: The American Aviation and Aerospace Industry.* Washington, DC: Smithsonian Institution Press, 2001.

———. "Putting Aircraft to Work: The First Air Freight." *Ohio History* 76, no. 4 (Autumn 1967): 247–58.

"Bonney Goes to Fly in Mexico." *Aerial Age Weekly* 1, no. 6 (26 April 1915): 128.

Bowers, Peter M. *Curtiss Aircraft, 1907–1947.* Annapolis: Naval Institute Press, 1979.

Boyer, Chip. "The Door: A Narrative and Commentary Concerning the Funerals of the Wright Brothers and of Other Wright Family Members, 1912–1948." Dayton: n.p., 2002.

Broadberry, Stephen N., and Douglas A. Irwin. "Labour Productivity in the United States and the United Kingdom during the Nineteenth Century." Discussion Paper, Centre for Economic Policy Research, London, 2005.

Brunsman, Charlotte K., and August E. Brunsman. *The Other Career of Wilbur and Orville: Wright & Wright Printers.* Kettering, OH: Trailside Press, 1989.

"The Burgess Company Expanding." *Aerial Age Weekly* 3, no. 2 (27 March 1916): 58.

"The Burgess Flying Boat Built for Robert J. Collier." *Aircraft* 4, no. 7 (September 1913): 164.

Cajka, Peter S. "The National Cash Register Company and the Neighborhoods: New Perspectives on Relief in the Dayton Flood of 1913." *Ohio History* 118 (2011): 49–71.

"The Camera Looks at Inland's Progress. . . ." *Inlander,* Special 15th Anniversary Edition (1937): 8.

Carroll, Francis M. *The American Presence in Ulster: A Diplomatic History, 1796–1996.* Washington, DC: Catholic University of America Press, 2005.

Chadeau, Emmanuel. *De Blériot à Dassault: Histoire de l'industrie aéronautique en France, 1900–1950.* Paris: A. Fayard, 1987.

Chandler, Charles de Forest, and Frank P. Lahm. *How Our Army Grew Wings: Airmen and Aircraft before 1914.* New York: Ronald Press, 1943.

"Court of Public Opinion." *Aeronautics* 12, no. 3 (March 1913): 90–91.

Crouch, Tom D. *The Bishop's Boys: A Life of Wilbur and Orville Wright.* New York: Norton, 1989.

————. "Chauffeur of the Skies: Roy Knabenshue and the Gasbag Era." *Timeline* 28, no. 2 (April–June 2011): 24–41.

————. *Wings: A History of Aviation from Kites to the Space Age.* New York: Norton, 2003.

Curtiss, Glenn Hammond, and Augustus Post. *The Curtiss Aviation Book.* New York: Frederick A. Stokes Co., 1912.

"The Curtiss Aeroplane and Motor Company Acquires Services of W. Starling Burgess and Burgess Company." *Aerial Age Weekly* 2, no. 22 (14 February 1916): 519.

Deines, Ann. "Roy Knabenshue: From Dirigibles to NPS." *CRM* 23, no. 2 (2000): 20–21.

"Description of the Burgess-Dunne Hydro-aeroplane." *Aircraft* 5, no. 2 (April 1914): 296–98.

Dienstbach, Carl. "The Rise of the Flying Machine Industry in America." *American Aeronaut* 1, no. 1 (August 1909): 12.

Douglas, Deborah G., and Amy E. Foster. *American Women and Flight since 1940.* Lexington: University Press of Kentucky, 2004.

Drury, Augustus W. *History of the City of Dayton and Montgomery County, Ohio.* Chicago: S. J. Clarke, 1909.

Duckworth, J. Herbert. "Aviation Happenings at Home." *Town and Country* 66, no. 1 (27 May 1911): 54–55.

Eckert, Michael. "Strategic Internationalism and the Transfer of Technical Knowledge: The United States, Germany, and Aerodynamics after World War I." *Technology and Culture* 46, no. 1 (January 2005): 104–31.

Edwards, John Carver. *Orville's Aviators: Outstanding Alumni of the Wright Flying School, 1910–1916.* Jefferson, NC: McFarland, 2009.

Eisenhower, Dwight D. "Report on Trans-Continental Trip," 3 November 1919. www .fhwa.dot.gov/infrastructure/convoy.cfm.

"Elton Aviation Company Organized." *Aeronautics* 15, no. 7 (15 October 1914): 104.

Eltscher, Louis, and Edward M. Young. *Curtiss-Wright: Greatness and Decline.* New York: Twayne, 1998.

"Entire World Mourns the Loss of Wilbur Wright." *Fly Magazine* 4, no. 9 (July 1912): 8–10.

Fahey, James C. *U.S. Army Aircraft (Heavier-than-Air) 1908–1946.* Falls Church, VA: Ships and Aircraft, 1946.

Faroute, Fay L., ed. *Aircraft Year Book, 1919.* New York: Manufacturers Aircraft Association, 1919.

"Fifteen Aeroplanes a Day for Europe." *Aerial Age Weekly* 2, no. 3 (4 October 1915): 53.

Freudenthal, Elsbeth E. *The Aviation Business: From Kitty Hawk to Wall Street.* New York: Vanguard, 1940.

Friedel, Robert. *Zipper: An Exploration in Novelty.* New York: Norton, 1994.

Galison, Peter, and Alex Roland, eds. *Atmospheric Flight in the Twentieth Century.* Archimedes, vol. 3: New Studies in the History and Philosophy of Science and Technology. Dordrecht: Kluwer Academic, 2000.

Gantenbein, Douglas. "Aviation's Birth Certificate." *Air and Space Magazine* 17, no. 6 (March 2003): 78–79.

Gerber, David A. *Black Ohio and the Color Line, 1860–1915.* Urbana: University of Illinois Press, 1976.

Gibson, Campbell. "Population of the 100 Largest Cities and Other Urban Places in the United States: 1790 to 1990." Population Division Working Paper 27. Washington, DC: U.S. Bureau of the Census, Population Division, 1998. www.census.gov/population /www/documentation/twps0027/twps0027.html.

Glenshaw, Paul. "Kings of the Air." *Air and Space Magazine* 27, no. 7 (February–March 2013): 46–51.

———. "Ladies and Gentlemen: The Aeroplane!" *Air and Space Magazine* 23, no. 1 (April–May 2008): 48–55.

Gould, Bartlett. "Burgess of Marblehead." *Essex Institute Historical Collections* 106, no. 1 (January 1970): 3–31.

Hallion, Richard P. "The Wright Kites, Gliders, and Airplanes: A Reference Guide." 19 August 2003. www.af.mil/shared/media/document/AFD-051013-002.pdf.

Harvard College. Class of 1901. *Secretary's Third Report, June, 1911*. Cambridge, MA: Crimson Printing Company, 1911.

Hayward, Charles B. *Practical Aeronautics: An Understandable Presentation of Interesting and Essential Facts in Aeronautical Science*. Chicago: American Technical Society, 1918.

Hennessy, Juliette. *The United States Army Air Arm, April 1861 to April 1917*. 1958. Washington, DC: Office of Air Force History, U.S. Air Force, 1985.

Henry, Lyell D., Jr. *Zig-Zag-and-Swirl: Alfred W. Lawson's Quest for Greatness*. Iowa City: University of Iowa Press, 1991.

Hounshell, David A. *From the American System to Mass Production, 1800–1932: The Development of Manufacturing Technology in the United States*. Baltimore: Johns Hopkins University Press, 1984.

Howard, Fred. *Wilbur and Orville: A Biography of the Wright Brothers*. New York: Knopf, 1987.

Interborough Rapid Transit Company. *Interborough Rapid Transit: The New York Subway; Its Construction and Equipment*. 1904. Reprint, New York: Arno, 1969.

"It Certainly Is a Darling." *Motor West* 27, no. 2 (1 May 1917): 26.

Jane, Fred T., ed. *Jane's All the World's Aircraft, 1913*. London: Sampson Low, Marston, 1913. Reprint, New York: Arco, 1969.

———. *Jane's All the World's Airships, 1909*. London: Sampson Low, Marston, 1909. Reprint, New York: Arco, 1969.

Johnson, Herbert A. *Wingless Eagle: U.S. Army Aviation through World War I*. Chapel Hill: University of North Carolina Press, 2001.

———. "The Wright Patent Wars and Early American Aviation." *Journal of Air Law and Commerce* 69, no. 1 (Winter 2004): 21–63.

Jones, Ernest. "A Review of 1911—Forecast for 1912." *Aeronautics* 10, no. 1 (January 1912): 1–4.

Judd, Richard W. *Socialist Cities: Municipal Politics and the Grass Roots of American Socialism*. Albany: State University of New York Press, 1989.

Keefer, Kathryn A. "The Wright Cycle Shop Historical Report: Greenfield Village," EI 186, Wright Cycle Shop—Simmons Intern Project 2005, Benson Ford Research Center, Dearborn, Michigan.

Kelly, Fred C. *The Wright Brothers: The Authorized Biography of Two Americans Whose Inventive Genius Changed the World*. 2nd ed. New York: 1943; Farrar, Straus, and Young, 1950.

Kirby, John, Jr., *The Career of a Family; or, the Ups and Downs of a Lifetime.* Dayton: Otterbein Press, 1916.

Knappen, Theodore. *Wings of War: An Account of the Important Contribution of the United States to Aircraft Invention, Engineering, Development and Production during the World War.* New York: G. P. Putnam's Sons, 1920.

Lebow, Eileen F. *Cal Rodgers and the* Vin Fiz: *The First Transcontinental Flight.* Washington, DC: Smithsonian Institution Press, 1989.

Leslie, Stuart W. *Boss Kettering.* New York: Columbia University Press, 1983.

Loening, Grover C. "The New Wright Aeroboat Type 'G.'" *Aeronautics* 14, no. 11 (15 June 1914): 171.

———. "The New Wright Model 'E' Single Propeller Biplane and the New Wright Six-Cylinder Motor." *Aircraft* 4, no. 9 (November 1913): 210.

———. *Our Wings Grow Faster: In These Personal Episodes of a Lifetime in Aviation May Be Found an Historical and Pictorial Record Showing How We So Quickly Stepped into This Air Age—and through What Kinds of Difficulties and Developments We Had to Pass to Get There.* Garden City, NY: Doubleday, Doran, 1935.

———. *Takeoff into Greatness: How American Aviation Grew So Big So Fast.* New York: Putnam, 1968.

———. "The Wright Company's New Hydro-Aeroplane Model C-H." *Aircraft* 4, no. 7 (September 1913): 152–53.

"Machinery Markets and News: The Central South." *Iron Age* 100, no. 22 (29 November 1917): 1344–45.

Mansfield, Howard. *Skylark: The Life, Lies, and Inventions of Harry Atwood.* Hanover, NH: University Press of New England, 1999.

Manufacturers Aircraft Association. *Aircraft Year Book, 1920.* New York: Doubleday, Page and Company, 1920.

"Manufacturers Make Product Suggestions." *Aviation and Aeronautical Engineering* 4, no. 11 (1 July 1918): 781.

Massachusetts. Bureau of Statistics, and Charles F. Gettemy. *The Decennial Census, 1915.* Boston: Wright and Potter, 1918.

McFadden, Thomas W. "Building Industries: Collective Action Problems and Institutional Solutions in the Development of the U.S. Aviation Industry, 1903–1938." PhD dissertation, University of Arizona, 1999.

Meyer, Stephen. "Adapting the Immigrant to the Line: Americanization in the Ford Factory, 1914–1921." *Journal of Social History* 14, no. 1 (Autumn 1980): 67–82.

Miller, Warren H. "The Manufacture of French Aeroplanes." *Engineering Magazine* 39, no. 5 (August 1910): 649.

Mitchell, Fred. "Historic and Architectural Resources of the Webster Station Area, Dayton, Ohio." National Register of Historic Places Multiple Property Documentation Form, 2000.

"The Model 'CH' Wright Waterplane." *Flight* 5, no. 36 (6 September 1913): 978.

Modley, Rudolf, and Aerospace Industries Association of America. *Aviation Facts and Figures, 1945.* New York: McGraw-Hill, 1945.

Morris, Lloyd. *Ceiling Unlimited: The Story of American Aviation from Kitty Hawk to Supersonics.* New York: Macmillan, 1953.

Morrow, John H. *The Great War in the Air: Military Aviation from 1909 to 1921.* Washington, DC: Smithsonian Institution Press, 1993.

"Motorist Buys Wright Plane." *Aero* 1, no. 24 (18 March 1911): 202.

"Navy Officer Trains at Wright School." *Aeronautics* 8, no. 5 (May 1911): 159.

"Navy Opens Bids for Nine Hydros." *Aeronautics* 16, no. 1 (15 March 1915): 4–5.

"Nearly 200 Winter Students Are Booked." *Aero and Hydro* 5, no. 12 (21 December 1912): 209.

Nelson, Daniel. *Farm and Factory: Workers in the Midwest, 1880–1990.* Bloomington: Indiana University Press, 1995.

———. "The New Factory System and the Unions: The National Cash Register Dispute of 1901." *Labor History* 15, no. 2 (1974): 163–78.

"New American Endurance Record." *Aeronautics* 9, no. 5 (November 1911): 175.

"New Companies." *Aeronautics* 10, no. 5 (May–June 1912): 178–79.

"The New Wright Biplane." *Flight* 6, no. 47 (20 November 1914): 1133.

"The New Wright Control." *Flight* 6, no. 10 (7 March 1914): 240.

"The New Wright Tractor Biplane—Type L." *Flight* 8, no. 32 (10 August 1916): 664.

"Night and Day Bank Building." *United States Investor* 20, no. 43 (6 November 1909): 1986.

Olszowka, John S. "From Shop Floor to Flight: Work and Labor in the Aircraft Industry, 1908–1945." PhD dissertation, Binghamton University, 2000.

"The 100 H.P. Curtiss Flying Boat." *Flying* 6, no. 15 (11 April 1914): 384–87.

Pattillo, Donald M. *Pushing the Envelope: The American Aircraft Industry.* Ann Arbor: University of Michigan Press, 1998.

Pulsifer, Woodbury, ed. *Navy Yearbook.* Washington, DC: Government Printing Office, 1911.

Robie, Bill. *For the Greatest Achievement: A History of the Aero Club of America and the National Aeronautic Association.* Washington, DC: Smithsonian Institution Press, 1993.

Robischon, E. W. "The Evolution of the Wright Airplane." *Aerosphere* 4 (1943): 129.

Roland, Alex. *Model Research: The National Advisory Committee for Aeronautics, 1915–1958.* Washington, DC: Scientific and Technical Information Branch, National Aeronautics and Space Administration, 1985.

Roseberry, C. R. *Glenn Curtiss: Pioneer of Flight.* 1972. Reprint, Syracuse, NY: Syracuse University Press, 1991.

Russell, Frank H. "Burgess Flying Boat Solves Warping Difficulties." *Aircraft* 4, no. 4 (June 1913): 79.

Scharff, Robert, and Walter S. Taylor. *Over Land and Sea: A Biography of Glenn Hammond Curtiss.* New York: D. McKay, 1968.

Scranton, Philip. "Diversity in Diversity: Flexible Production and American Industrialization, 1880–1930." *Business History Review* 65, no. 1 (Spring 1991): 27–90.

Sealander, Judith. *Grand Plans: Business Progressivism and Social Change in Ohio's Miami Valley, 1890–1929.* Lexington: University Press of Kentucky, 1988.

Seelhorst, Mary. "Popular Mechanics: 90 Years." *Popular Mechanics* 169, no. 2 (February 1992): 83–85.

Shulman, Seth. *Unlocking the Sky: Glenn Hammond Curtiss and the Race to Invent the Airplane.* New York: HarperCollins, 2002.

Bibliography

"Simplex Motor with Wright Aeroplane Co." *Aerial Age Weekly* 2, no. 11 (29 November 1915): 248.

"The 60 H.P. Wright Aero-Boat." *Flying* 6, no. 2 (17 January 1914): 56–58.

Strother, French. "Flying across the Continent: Rodgers' Trip from Kansas City to Pasadena." *World's Work* 23, no. 4 (February 1912): 399–408.

Swanborough, Gordon, and Peter M. Bowers. *United States Navy Aircraft since 1911.* Annapolis: Naval Institute Press, 1990.

Thum, Robert. "The First Jewish Aviator." *Dayton Jewish Observer* 7, no. 6 (January 2003): 18–19.

Trimble, William F. *High Frontier: A History of Aeronautics in Pennsylvania.* Pittsburgh: University of Pittsburgh Press, 1982.

Trostel, Scott D. *The Barney and Smith Car Company: Car Builders, Dayton, Ohio.* Fletcher, OH: Cam-Tech Publishing, 1993.

U.S. Department of Commerce. Economics and Statistics Administration. Bureau of the Census. *Statistical Abstract of the United States: 2003.* Washington, DC: Government Printing Office, 2003. www.census.gov/statab/hist/HS-22.pdf.

U.S. Department of Commerce. Bureau of the Census. *Thirteenth Census of the United States Taken in the Year 1910.* Vol. 3, *Population—Reports by States, Nebraska-Wyoming, Alaska, Hawaii and Porto Rico.* Washington, DC: Government Printing Office, 1913.

———. *Thirteenth Census of the United States Taken in the Year 1910.* Vol. 4, *Population: Occupational Statistics.* Washington, DC: Government Printing Office, 1913.

U.S. Department of Commerce and Labor. Bureau of Foreign and Domestic Commerce. *Statistical Abstract of the United States 1912, Thirty-fifth Number.* Washington, DC: Government Printing Office, 1913.

U.S. Patent Office. *Annual Report of the Commissioner of Patents for the Year 1915.* Washington, DC: Government Printing Office, 1916.

Walker, John T. "Socialism in Dayton, Ohio, 1912 to 1925: Its Membership, Organization, and Demise." *Labor History* 26, no. 3 (Summer 1985): 384–404.

Warnock, A. Timothy. "From Infant Technology to Obsolescence: The Wright Brothers' Airplane in the U.S. Army Signal Corps, 1905–1915." *Air Power History* 49, no. 4 (Winter 2002): 46–57.

"Who's Who in Aeronautics." *Aerial Age Weekly* 13, no. 3 (28 March 1921): 61.

Williams, Julie Hedgepeth. *Wings of Opportunity: The Wright Brothers in Montgomery, Alabama, 1910; America's First Civilian Flying School and the City That Capitalized on It.* Montgomery: NewSouth Books, 2010.

Williams Directory Company. *Williams' Dayton Directory for 1910–1911.* Cincinnati: Williams Directory Company, 1910.

———. *Williams' Dayton Directory for 1911–1912.* Cincinnati: Williams Directory Company, 1911.

———. *Williams' Dayton Directory 1912–1913.* Cincinnati: Williams Directory Company, 1912.

———. *Williams' Dayton Directory 1913–1914.* Cincinnati: Williams Directory Company, 1913.

———. *Williams' Dayton Directory 1914–1915.* Cincinnati: Williams Directory Company, 1914.

———. *Williams' Dayton Directory 1915–1916.* Cincinnati: Williams Directory Company, 1915.

Wright, Horace. "Recollections." In *Wright Reminiscences,* edited by Ivonette Wright Miller, 156–62. Wright-Patterson Air Force Base, OH: Air Force Museum Foundation, 1978.

Wright, Milton. *Diaries, 1857–1917.* Dayton: Wright State University, 1999.

Wright, Wilbur, and Orville Wright. *Miracle at Kitty Hawk: The Letters of Wilbur and Orville Wright.* Edited by Fred C. Kelly. New York: Farrar, Straus and Young, 1951. Reprint, New York: Da Capo, 1996.

Wright, Wilbur, Orville Wright, and Octave Chanute. *The Papers of Wilbur and Orville Wright, Including the Chanute-Wright Letters and Other Papers of Octave Chanute.* Edited by Marvin W. McFarland. New York: McGraw-Hill, 1953. Reprint, Salem, NH: Ayer, 1990.

Wright Company. *The Wright Flyer.* Dayton: United Brethren Publishing House, 1911.

———. *Wright Flyers.* New York: Premier Press, 1912.

"Wright Company, New York, and Glenn L. Martin Company, Los Angeles, Merge." *Aerial Age Weekly* 3, no. 22 (14 August 1916): 653.

"Wright Company's Winter School in the South." *Aerial Age Weekly* 2, no. 14 (20 December 1915): 327.

"Wright-Curtiss Litigation." *Aeronautics* 12, no. 3 (March 1913): 85.

"Wright Licenses Granted." *Aeronautics* 14, no. 10 (30 May 1914): 147.

"Wright News." *Aerial Age Weekly* 1, no. 14 (21 June 1915): 319.

Yates, JoAnne. *Control through Communication: The Rise of System in American Management.* Baltimore: Johns Hopkins University Press, 1989.

# INDEX

An illustration is indicated by a page number in italic type. The letter *n* following a page number indicates a note on that page. The number following the *n* is the note number.

CPSIA information can be obtained at www.ICGtesting.com
Printed in the USA
LVOW11*2315021014

407069LV00002B/2/P